Industrial Wastewater Source Control

AN INSPECTION GUIDE

NANCY RIIKONEN, Author, *City of San Diego*
CLAY JONES, Instructional Designer, *Redwood Instructional Services*

TECHNOMIC PUBLISHING CO., INC.
LANCASTER · BASEL

Industrial Wastewater Source Control
a TECHNOMIC publication

Published in the Western Hemisphere by
Technomic Publishing Company, Inc.
851 New Holland Avenue
Box 3535
Lancaster, Pennsylvania 17604 U.S.A.

Distributed in the Rest of the World by
Technomic Publishing AG

Copyright © 1992 by Technomic Publishing Company, Inc.
All rights reserved

No part of this publication may be reproduced, stored in a
retrieval system, or transmitted, in any form or by any means,
electronic, mechanical, photocopying, recording, or otherwise,
without the prior written permission of the publisher.

Printed in the United States of America
10 9 8 7 6 5 4 3 2 1

Main entry under title:
    Industrial Wastewater Source Control: An Inspection Guide

A Technomic Publishing Company book
Bibliography: p. Bibliography-1
Includes index p. Index-1

Library of Congress Card No. 91-66930
ISBN No. 87762-855-6

HOW TO ORDER THIS BOOK
BY PHONE: 800-233-9936 or 717-291-5609, 8AM–5PM Eastern Time
BY FAX: 717-295-4538
BY MAIL: Order Department
Technomic Publishing Company, Inc.
851 New Holland Avenue, Box 3535
Lancaster, PA 17604, U.S.A.
BY CREDIT CARD: American Express, VISA, MasterCard

Table of Contents

Foreword v

Acknowledgements vii

User's Guide ix

Automotive Repair/Service A-1
Food Processing B-1
Hospitals C-1
Laboratories D-1
Laundries E-1
Paint and Ink Formulation F-1
Pharmaceuticals G-1
Photo Finishing H-1
Printing/Publishing I-1
Steam Plants/Boilers J-1
Large Institutions K-1
Metal Finishing/Electroplating L-1
Treatment Technologies M-1
Oil Recovery/Oily Waste Pretreatment N-1
Cathode Ray Tubes O-1
Semiconductors P-1
Cooling Systems Q-1

Glossary Glossary-1

Dictionary of Acronyms Acronyms-1

Bibliography Bibliography-1

Index Index-1

Foreword

This Inspection Guide is a compilation of standard methods, "know-how", data and investigation techniques which have proven valuable for industrial facility inspections. They have been "field tested" by City of San Diego industrial waste inspectors over several years in all kinds of industry facilities. The organization of this guide into instructional modules is designed to make this a valuable tool for training new inspectors and as a quick reference for "old hands".

The Guide was originally conceived and prepared for the exclusive use of the City's metro liquid industrial waste program because after years of diligent and discouraging searching, it became apparent that no such similar "hands-on .... This-is-how-you-do-it" manual was available in the marketplace.

With all due humility, this manual, as designed by its editor, is intended to be a complete and comprehensive guide for industrial waste inspectors, while providing a simplified, no-nonsense, illustrated text for a basic understanding of a vastly complex field for both the novice and the experienced inspector-technician. A comprehensive Bibliography will direct Guide users to supplementary information sources to solve "special" problems. The Glossary clarifies and defines all of the specialized technical terms used in the text.

Because San Diego's industrial facilities are so varied in size and function, the City and Technomic Publishing Company with the City's Engineer-Consultant, Hirsch & Company, have undertaken to reproduce the Guide in order to make its contents available nationally to what is believed to be a broad based user group with similar interests and needs.

The user, we hope, will gratefully discover that the Guide has been designed into an especially easy reading format, with ample illustrations and "white space" for notes. The use of instructional design principles guided these aspects of the manual. Hopefully, it will be carried into the field and will serve as a "technical bible". The Guide is intended to be a "living" document and may be upgraded by the user or publisher with new or more comprehensive topic "modules" from time to time.

In this first edition of the Guide, there are 16 sections or "modules", each self-contained and dealing with a specific topic:

Automotive Repair/Service	Printing/Publishing
Food Processing	Steam Plants/Boilers
Hospitals	Large Institutions
Laboratories	Metal Finishing/Electroplating
Laundries	Cathode Ray Tubes
Paint and Ink Formulation	Semiconductors
Pharmaceuticals	Treatment Technologies
Photo Finishing	Oil Recovery/Oily Waste Pretreatment

The appendices offer the user information as to various Cooling Systems and Engineer-Estimating hydraulic flow and consumption data.

The manual can be adopted in whole or in part, or amended to serve the needs of any agency concerned with field level inspections and regulation of industrial/commercial wastewater pollutant production, control and discharge.

The manual should also serve as a unique training aid and guide for field level/technical dischargers and operators of industrial/commercial wastewater pollutant control systems.

In developing this Guide, the Author/Editor sorted and synthesized literally mountains of EPA/federal technical data. Ms. Riikonen is an experienced industrial waste inspector and currently heads the City of San Diego's Compliance Section. Mr. Jones is an experienced instructional designer with an advanced degree in Educational Technology. Hirsch & Company is a San Diego based consulting engineering firm specializing in water and wastes engineering for over twenty years. Coordination of this manual for Hirsch & Company was under the direction of the firm's principal, Lawrence Hirsch, a registered professional engineer with over thirty years of water pollution abatement experience.

In short, the emphasis is on practical, up-to-date information and guidance to meet the daily needs of the field inspector and to serve as a text for an inspectors training program or in-plant operator guidance, organized in a unique format, filling a void in the instructional tools available at the field level.

Acknowledgements

A collective thank you is in order for all the help and inspiration I received from many, many people during the researching and writing of this document.

To all of the inspectors; Susan Otten, Barbara Sharatz, André Smith, and John Steger; who diligently plodded through the early prototype modules and provided the author with necessary feedback, my warmest thanks. Also, special kudos go to Barbara Sharatz for her technical research and assistance on the Photo Finishing, Laundry, and especially the Pharmaceuticals module which is included herein solely due to her efforts. To John Steger, a special tribute for the care that went into producing first-class engineering drawings, irrespective of the various sketches, directions or impressions he had to work from.

It goes without saying that Clay Jones, Instructional Developer for Hirsch and Company, who worked with me throughout the duration of the project is one of a kind. His patience, perseverance, and meticulous attention to detail raised the level of this document from mundane manual to a vital instructional tool.

-NCR

User's Guide

Any attempt to complete a work of instruction, or reference, or a guide such as this is fraught with certain hazards. There are the questions that cannot be answered - is this too much information? Not enough? Is the information current? Helpful? Presented in a format that the inspector will find useful? We have tried to address these questions throughout the guide, and hope the results reflect favorably.

This guide is intended to be a <u>living</u> document. We hope it will be copied and carried into the field, dog-eared in places and the margins filled with notes.

It was toward this concept of the Inspection Guide as a working tool that certain design decisions were made. Those included providing ample "white space" on the pages, not only to assist readability, but to allow room for notes; dividing the subjects covered into modules, each numbered separately; and designing the total document for use in a three ring binder to enable removal of selected pages for copying and to carry into the field. To better assist you in using this guide, we offer the following "road map."

1. Each section, or module, carries the title of the industry it addresses.

2. Each module begins with a brief introduction, or overview, of the industry covered. This introduction concludes with a listing of subtopic areas within the module.

3. Subtopic areas within modules are represented by the symbol ❏.

4. Within Subtopic areas, information is further broken down by the use of bullets (•) and deltas (∆). Bullets usually indicate key points, deltas frequently represent lists.

5. Engineering illustrations, charts and tables are numbered according to the module in which they appear as well as the text page which refers to them.

So...the informational hierarchy within each module is represented as follows:

INTRODUCTION (The industry)

    ❑ (Major topics)

        • (Subtopics)

            Δ (Lists or points within subtopics)

Finally, each module and, frequently, subtopics within a module contain **Inspector's Questions.** These lists of questions are intended to be taken to the field by the user and represent, to the best of our ability, an overview of the most basic, necessary information which needs to be collected by the inspector at the type of industry for which the questions were written. They are certainly not all-inclusive, and many of the same questions will be asked at varying industries. However, we hope they serve as a useful guide.

- CMJ & NCR

Section A

AUTOMOTIVE REPAIR/SERVICES

Introduction

Although the automotive repair industry encompasses many aspects, this section will focus on those shops which could or do generate some type of industrial wastewater subject to local regulations. The automotive facilities which will be examined in this section are the following (and include related Inspector's Questions for each):

- **New Car Dealerships/Auto Rental Agencies**

- **Automotive Repair Shops**

    - **Machine Shops**

    - **Engine and/or Transmission Repair**

    - **Radiator Service**

    - **Steam Cleaning/High Temperature-High Pressure Washing**

- **Car Washes**

- **Paint Shops/Body Shops/Detail Shops**

❑ New Car Dealerships

New car dealerships and auto rental agencies represent a sizeable share of the automotive discharger category. However, the nature of their discharge effluent is relatively innocuous because they are usually not involved in the grosser aspects of auto repair. This is not to say, however, that a poorly run or managed facility could not contribute excessive amounts of automotive oil, steam cleaning sludge or even dangerous flammable solvents to the sewer system.

The most common permitted operation at these firms is the car wash. This is usually set up in a wash bay, and fully equipped with a small clarifier, single or double chambered, with a suitable plumber's elbow at the discharge side. Sometimes de-ionized water is used but usually City water is in use. Generally there is no recycling of water due to the relatively low volume of water used.

Other permitted operations which could be present are steam cleaning or high pressure-high temperature washing and new car prep operations (often using and discharging prohibited chemicals).

It is not uncommon to find that daily washing of all cars occurs. This daily washing consists of a light application of de-ionized water from a portable car washing unit. No soap is used, as in the car wash bay, and the flow is generally of a low enough volume that the discharge never leaves the car lot. Occasionally, the inspector will encounter a dealership or a rental agency which does _all_ of its car washing without benefit of sewer hookup. This presents a problem that must be addressed. Specifically, the owner/operator must either stop discharging (to such an extent that no water leaves their property), or they can connect to the sanitary sewer via conventional plumbing structures.

INSPECTOR'S QUESTIONS—NEW CAR DEALERSHIPS/RENTAL AGENCIES

1. What types of water use operations are present at this facility?

    - Car Wash (Bay)?
    - Car Wash (Outside)?
    - New Car Prep?
    - Parts Cleaning?
    - Steam Cleaning?
    - High Temperature-High Pressure Washing?
    - Other?

2. Is any degreasing done? Describe...

3. Are alkaline cleaning or acid cleaning tanks present?
    - If yes, how frequently are these tanks batch discharged to the sewer?
    - What pretreatment occurs prior to discharge?
    - What are the individual volumes of all process tanks in facility?
    - How is bottom sludge handled?

4. How are spent or contaminated processing tanks handled?
    - If hauled, who hauls it? (Can check manifests.)

5. What types of repair or maintenance work is done?
    - Location of work?
    - Are there any sewer or storm drains in the area?

6. Any rectifiers, compressors or similar equipment in use?
    - If yes, do any of them require cooling water?
    - Any one-pass cooling water?
    - If yes, volume (gallons per day)?
    - Where is the discharge point?

7. Is there any routine washdown of the work area(s)?

- Where is the discharge point?
- Is a grease and oil interceptor present?

8. Are there any floor drains in the process area?
    - Does their flow pass through a common interceptor point?
    - If no, where does it discharge?

9. Do all process streams flow to a common interceptor point?
    - If yes, where is that point?
    - If no, where are the various points?

10. Is there a grease interceptor that receives the wastewater?
    - If yes, what is the size of the interceptor?
    - Where is it located?
    - Is an elbow present at the discharge side?
    - If no, explain why absent?
    - Is this discharge properly routed to sewer? Explain...

11. Is there any pretreatment of waste water?
    - If yes, describe the various aspects in place and where located.

12. How is the routine disposal of hazardous wastes handled?
    - Who is the waste hauler?
    - What is the average volume of material disposed to landfill, recycled, etc.?
    - Where are records kept? (Can ask to see manifests.)

13. Is there any de-ionized water used on site?
    - If yes, is it on a service? Who maintains the equipment?
    - What, if any, pretreatment is there for the systems?...Describe?
    - Is water kept in a holding tank and drawn off as needed?
    - Or is it produced on an as-needed basis?
    - Is the water metered in? Out?

14. Any requirements for steam generation or use?
    - If yes, is water metered?
    - Volume (gallons per day)?

15. Is there any water reuse?
    - If yes, where and what is it?

16. What solvents are used at this facility?
    - Where are they used? What unit operations?
    - What is the method of application? Dip tank, rag applied, or other?
    - Is there any discharge of solvents to sewer in any form? From the unit operation(s)? Final triple rinsing of containers prior to disposal? Explain...
    - If no solvents are discharged to sewer, explain what procedures are in place to prevent discharge. Solvent Management Plan (SMP)?
    - When was SMP submitted? Explain...

17. Is there adequate spill containment in the processing area(s)?
    - If yes, describe.
    - If no, describe.
    - How is the sewer otherwise protected from leaks or spills in the process area(s) and throughout the facility? Describe in detail.

❏ Automotive Repair Shops

Of all the automotive categories which discharge to the sewer, this group is the most problematic. They are the grease and oil dischargers and the heavy metals dischargers. Many of them also use chemicals which can cause their discharge pH to suffer wide swings on both sides of the permitted limits. Individually, their flows do not usually exceed 1,000 gallons per day, but collectively they can represent a meaningful contribution to the collection system.

• **Machine Shops** are primarily involved with specialized manufacturing of metal parts for all types of machinery, of which automotive is but a single activity. The most common waste water discharge originates from non-contact cooling water* for spot welders, air compressors, vapor degreasers, and milling machines. Occasionally, cooling towers may be present but this is not likely in the small shop.

Many varieties of cutting oils and water-based coolants are used in machining and working with the various metals, but these materials are not normally discharged to the sewer except perhaps in the smaller shops. The exceptions are water-based coolants and cutting oils. In the larger shops all of the oils and coolants are collected and recycled because it is economically advantageous in addition to being disallowed by most sewer ordinances. The most common oils found in machine shops are:

△ Hydraulic fluids: only found in wastewater if hydraulic lines are changed or if lines rupture.

△ Lubricating oils: rarely found in wastewater. They are reused until consumed, or recycled to an oil reclaimer.

△ Quenching oils: used in heat treatment of metal and not discharged to wastewater.

△ Cutting oils: used as coolants in machining and are commonly found in wastewater. Can be of the insoluble type or soluble type. Cutting oil must be changed with a regular frequency because the oil eventually breaks down, becomes rancid, and is unfit for use. Small amounts are

---

* Most municipal ordinances have a prohibition against the discharge of single-pass or non-contact cooling water to the sanitary sewer.

probably routinely discharged to sewers with larger volumes either being hauled away by disposal services or recycled/reclaimed on site.

Also, acidic or caustic solutions are occasionally used to clean metal parts. Usually there are post-cleaning rinses associated with these acids and caustics. They may be "dead rinses," running rinses, or some combination of the two. Both the acid and caustic solutions have to be discharged when they become spent or contaminated. The caustic solutions generally lose their effectiveness more quickly and are discharged more frequently. In every case, the primary concerns are proper handling of any sludge residuals and that the liquid portion discharged to sewer meets the local pH limit .

INSPECTOR'S QUESTIONS—MACHINE SHOPS

1. What type of product is produced?

2. What type of material is machined?

3. What types of coolants and/or cutting oils are used?
    - Water soluble?
    - Insoluble?

4. Are these coolants/cutting oils ever discharged to sewer?
    - If yes, what frequency?
    - What volume?
    - Where does the discharge occur?
    - If no, how are spent or contaminated coolant materials handled?
    - Who is the hauler?
    - Are they recycled?
    - Where? On-site?

5. Is any degreasing done?
    - Caustic soak (hot tank)?
    - Vapor degreasing?
    - Safety Kleen?
    - Bake-off oven?
    - Jet spray?
    - Steam cleaner?

6. Is any vapor degreasing done?
    - If yes, is the unit water cooled (one-pass), on a cooling tower or refrigerant type?
    - If one-pass, where does it discharge to sewer? (One-pass not usually allowed by sewer ordinance.)
    - What type of solvent is used in the vapor degreaser?
    - How is the spent or contaminated solvent handled? Describe...

7. Is there any rinsing after the degreasing?
   - Still rinse?
   - Running rinse?
   - Combination?

8. Is there single-pass cooling water used for compressors, vapor degreasing, rectifiers, etc.? If yes, describe what it is used for?
   - If yes, what is the volume (gallons per day)?
   - Where is the discharge location?

9. Is there any routine washdown of the work area(s)?
   - Where is the discharge point?
   - Is a grease and oil interceptor present?

10. Are there any floor drains in the process area?
    - Does their flow pass through a common interceptor point?
    - If no, where does it discharge?

11. Do all process streams flow to a common interceptor point?
    - If yes, where is that point?
    - If no, where are the various points?
    - Is one-pass cooling water co-mingling with regulated wastewater at the sample point(s)?

12. Is there a grease interceptor that receives the wastewater?
    - If yes, what is the size of the interceptor?
    - Where is it located?
    - Is an elbow present at the discharge side?
    - If no, explain why absent?
    - Is this discharge properly routed to sewer? Explain...

13. Is there any pretreatment of wastewater?
    - If yes, describe the various aspects in place and where located.

14. How is the routine disposal of hazardous wastes handled?
    - Who is the waste hauler?
    - What is the average volume of material disposed to landfill, recycled,

etc.?
- Where are the records kept? (Can ask to see manifests.)

15. Any requirements for steam generation or use?
    - If yes, is make-up water metered?
    - Volume (gallons per day)?
    - Is there blowdown to sewer? ...Frequency? ...Volume (gallons per day)?
    - What are the steam losses (gallons per day)?

16. Is there any water reuse?
    - If yes, where and what is it?

17. Is there adequate spill containment in the process area(s)?
    - If yes, describe.
    - If no, describe.
    - How is the sewer otherwise protected from leaks or spills in the process area(s) and throughout the facility? Describe in detail.

Δ **Engine and transmission repair shops** can be very heavy contributors of grease and oil if no form of pretreatment is present. The exception is when degreasing is done exclusively with a "bake-off" method. All of the other degreasing methods; steam cleaning, jet spray, post solvent rinse or post-caustic (hot tank) rinse, used alone or in some combination with each other; all require some type of grease interceptor or clarifier in order to prevent wholesale discharge of grease and oil to the sewer system. Additionally, when a caustic soak (hot tank) is used, pH of the discharge effluent is a concern as is the proper handling of the tank bottoms (sludge).

INSPECTOR'S QUESTIONS—ENGINE/TRANSMISSION REPAIR

1. What type of repair activity is done?
    - Engine?
    - Transmission?
    - Other?

2. What type of degreasing is done?
    - Bake-off oven?
    - Safety Kleen Solvent?
    - Other type of solvent degreasing?
    - Jet spray?
    - Caustic Soak (hot tank)?
    - Steam cleaner?
    - Other?

3. Is any vapor degreasing done?
    - If yes, is the unit water cooled (one-pass), on a cooling tower or refrigerant type?
    - If one-pass, where does it discharge to sewer? (One-pass not usually allowed by sewer ordinance.)
    - What type of solvent is used in the vapor degreaser?
    - How is spent or contaminated solvent handled? Describe...

4. What are the individual volumes of the process tanks?
    - Describe...

5. Are alkaline cleaning or acid cleaning tanks present?
    - If yes, how frequently are these tanks batch discharged to the sewer?
    - What pretreatment occurs prior to discharge?
    - What are the individual volumes of all process tanks in facility?
    - How is bottom sludge handled?

6. How are spent or contaminated processing tanks handled?

- If hauled, who hauls it? (Can check manifests.)

7. Are any tanks heated?
    - If yes, which ones?

8. Any rectifiers, compressors or similar equipment in use?
    - If yes, do any of them require cooling water?
    - Any one-pass cooling water?
    - If yes, volume (gallons per day)?
    - Where is the discharge point?
    - If water is not used for cooling equipment, describe the cooling practices.

9. Is there any routine washdown of the work area(s)?
    - Where is the discharge point?
    - Is a grease and oil interceptor present?

10. Are there any floor drains in the process area?
    - Does their flow pass through a common interceptor point?
    - If no, where does it discharge?

11. Do all process streams flow to a common interceptor point?
    - If yes, where is that point?
    - If no, where are the various points?
    - Is one-pass cooling water co-mingling with regulated wastewater at the sample point(s)?

12. Is there any pretreatment of waste water?
    - If yes, describe the various aspects in place and where located.

13. Is there a grease interceptor that receives the wastewater?
    - If yes, what is the size of the interceptor?
    - Where is it located?
    - Is an elbow present at the discharge side?
    - If no, explain why absent?
    - Is this discharge properly routed to sewer? Explain...

14. Any requirements for steam generation or use?

-If yes, is make-up water metered?
-Volume (gallons per day)?
-Is there blowdown to sewer? ...Frequency? ...Volume (gallons per day)?
-What are the steam losses (gallons per day)?

15. Is there any de-ionized water use?
    -If yes, is it on a service? Who maintains the equipment?
    -What, if any, pretreatment is there for the system?...Describe?
    -Is water kept in a holding tank and drawn off as needed?
    -Or is it produced on an as-needed basis?
    -Is the water metered in? Out?

16. How is the routine disposal of hazardous wastes handled?
    - Who is the waste hauler?
    - What is the average volume of material disposed to landfill, recycled, etc.?
    - Where are the records kept? (Can ask to see manifests.)

17. Is there any water reuse?
    - If yes, where and what is it?

18. Is there adequate spill containment in the process area(s)?
    - If yes, describe.
    - If no, describe.
    - How is the sewer otherwise protected from leaks or spills in the process area(s) and throughout the facility? Describe in detail.

RADIATOR SHOPS

Radiator shops are an example of an industry type that generally has a low volume wastewater discharge. However, the concentration of the metals present can be extremely high, and the pH can be in the range of 2-12, depending on what operation is being performed.

The metals of concern are lead, zinc and copper. The following areas are covered in this module:

- ❏ Description of Process Area

- ❏ Origin and Destination of Metals in Wastewater

- ❏ Control of Metal in Wastewater

- ❏ Industry Conclusions

- ❏ Inspector's Questions

❑ Description of Process Area

The typical radiator shop is usually a small, independent facility or an adjunct operation to a larger full-service auto repair shop. Either way, the wastewater generated is unique and requires special attention because of the heavy load of pollutants which can be present. Most shops perform the following operations within the scope of radiator repair:

- Drain, flush and rod out radiators prior to fixing them
- Repair radiators using lead solder and a zinc-based flux
- Re-core radiators using lead solder and a zinc-based flux
- Test repaired radiators for leaks using a test tank containing fluorescine dye
- Flush and clean out newly repaired radiators
- Paint the repaired radiators

- Draining, Flushing and Rodding out of Radiators

It is at this initial stage of the repair operation that a certain amount of the heavy metals present in the old radiator fluid get introduced to the sewer system. As the radiator has aged, a considerable amount of internal metal degradation had taken place. When the radiator contents are drained, flushed, and/or rodded out this particulate metal is dislodged and discharged to sewer. Principally, the metals of interest are lead (from the solder), zinc (from the flux), and copper from the radiator itself. Iron, in the form of rust, is generally present in extensive amounts but is not a consideration, whereas the lead, zinc, and copper represent toxic heavy metals and are a cause for concern.

- Repairing the Radiator

The repair process consists of acid cleaning of the area to be repaired (with the acid flux) and running hot lead solder over the area of repair so that the hole is completely sealed and no leaking of radiator fluid is possible. It is common practice to do this work over the test tank. This is <u>not</u> a good procedure because any slopover of acid flux (zinc) or lead solder drops into the test tank: It is better to do all repairs over a dry container such as a 55-gallon drum. This more easily contains the metal residue until such time as proper disposal can be done.

- Re-coring

    The re-coring procedure generally consists of removing the top and bottom pans and replacing the damaged middle core with a new insert. The new core is reattached to the top and bottom pieces using the same solder-flux procedure as in the repair procedure. If re-coring is not needed, the old radiator is only "rodded out". This procedure removes the accumulated particulate metals from the radiator core in order to facilitate the flow of fluid through the cooling system.

- Testing the Radiator

    The testing procedure is done to check the integrity of the soldered area. The test tank contains a flourescine dye solution which is the leak detection medium. Essentially the presence of dye makes cracks or holes easier to see. The most common test tank chemical is called Thompson's Test Tank Powder and it produces a bright green liquid in the tank. The main problem the test tank poses is as an additional source of metals in the wastewater if the shop does repairing over the tank (and most do because of convenience). As the repair work continues over the tank, the test liquid becomes increasingly acidic (from the slopover of the acidic zinc flux) and must be discharged. In general, most shops have indicated they discharge their test tanks anywhere from once per week to once per month even if acidic conditions are not present because the test tanks simply become stagnant.

- Final Flushing of the Radiator

    The final flushing is done to remove the residual dye and clean up the radiator so it looks presentable to the customer. The pollutants of concern here are those constituents carried over from the test tank, namely; zinc, lead and copper, and conceivably low pH waste water.

- Painting

    Some shops may also refinish the final repaired product with a coat of black paint. There is normally no discharge of wastewater from this operation.

Typical Radiator Shop Layout
(Top View)

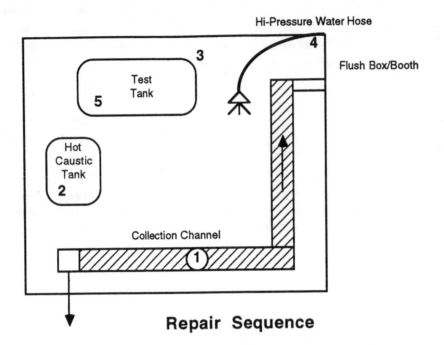

Repair Sequence

1. Drain Radiator (contains ethylene-glycol, zinc, copper and lead)

2. Clean Radiator in hot caustic/boilout tank (contains caustic, zinc, copper and lead - *not changed out except when spent)

3. Rod out Radiator

4. Flush Radiator

5. Repair Radiator

    - Repair damaged area with acid
    - Apply acid/zinc flux to area being repaired
    - Solder with lead solder

6. Place in test tank to check for leaks

7. Final Flush

* The bottom sludge always has to be handled as hazardous material-supernatant water can be discharged to sewer if pH and dissolved metals are within local sewer agency limits.

Figure A-1

❑ **Origin and Destination of Metals in Wastewater**

Origin	Pollutant	Destination
Draining/Flushing	Lead, Zinc & Copper	Sewer
Repairing/Rodding	Lead, Zinc & Copper	N/A - unless work done over test tank and tank dumped to sewer
Re-coring	Lead, Zinc & Copper	N/A - unless work done over test tank and tank dumped to sewer
Testing	Lead, Zinc & Copper	Sewer (only when tank is batch discharged)
Final Flushing	Lead, Zinc & Copper	Sewer

❑ Control of Metals in Wastewater

Metals are usually present in <u>two</u> forms in radiator shop discharges:

- Δ As particulate solids carried out by the water

- Δ As dissolved salts in the water

Removal of the heavy metals in the form of particles involves <u>one or both</u> of the following methods:

- Δ Settle out the solids in a sump or a similar device

- Δ Filter out the solids by some means

Removal of the dissolved solids requires that first the dissolved salts are converted into particulate solids, and then either of the two above methods for removal of particulate solids are applied.

For most of the metals involved in radiator shops, converting the dissolved metal to a solid particulate can be done by adding caustic in adequate amounts to a central collection sump or clarifier. The key to successful metals precipitation is to keep the wastewater caustic.

• Filtering/Settling

Settling used alone as a method may be too slow and excessive solids may be carried out by the wastewater flow. A filter at the outlet to the sewer could mitigate this situation. This type of application will usually require a pump or at least enough "head" to drain the water through a filter into the sewer connection. Settling can be enhanced by adding a small amount of material which "agglomerates" or aids the small particles in forming larger particles. Examples of this type of addition are alum, or corn starch. Addition of a 2-3% slurry of these materials will promote better settling.

As an alternative to a filter at the sewer connection, it might be practical to capture all wastewater in a small tank or clarifier. This water could be treated with caustic and agglomerates and allowed to settle overnight. The metal-free supernatant waste could then be drained to the sewer, and the settled solids collected for hazardous waste disposal.

Regardless of what type of metals removal system is employed, periodically all sumps or collection vats must be cleaned out to remove accumulated solids. (The initiation of the cleaning process in no way rescinds the imposed metal limits. Cleaning is considered a normal process operation and subject to the same limits as the radiator repairing itself.) If capturing the wastewater is not practiced, and the shop must discharge wastewater continuously, a sump with multiple chambers, or a series of small sumps, can be used to help with the settling process.

❑ Industry Conclusions

In general, it can be said that the radiator repair business is not a growth industry. For the most part, business is falling off in many areas because of increased environmental controls and the inability of existing shops to keep up with the costs of doing business. The shops are either unable to make their expenses because of the high costs associated with hauling away the hazardous wastes they generate, or they are unable or unwilling to convert to more environmentally sound and less water-intensive systems such as total recycling systems with no discharge to sewer. Additionally, many new automobiles are being built with plastic throwaway radiators. When the plastic radiator leaks or fails in some other way, a new model replaces the old one and repairs are not a consideration - only replacement.

In the not-too-distant future, radiator repair shops will be a minor industrial wastewater category, dealing primarily with specialty repairs on older model cars or other vehicles for which there is no plastic replacement radiator available.

INSPECTOR'S QUESTIONS—RADIATOR SHOPS

1. Describe the radiator repair operation as it exists at this shop.
    - Does it include re-coring of radiators?
    - Is there a boil-out tank? Explain...
    - Is there a test tank? Explain...

2. Is this a standard type of radiator operation with traditional applications, or is it a water conservative or "no discharge" type of system?
    - Explain...

3. Is there any water reuse within the shop?
    - If yes, where?
    - How is it occurring?
    - Explain...

4. Any rectifiers, compressors or similar equipment in use?
    - If yes, do any of them require cooling water?
    - Any one-pass cooling water?
    - If yes, volume (gallons per day)?
    - What is the discharge point?
    - If water is not used for cooling equipment, describe cooling practices.

5. Is there any routine washdown of the work area(s)?
    - Where is the discharge point?
    - Is an interceptor or separation sump present?

6. Are there any floor drains in the process area (other than the sample point)?
    - Does the flow from these other floor drains flow to a common interceptor point?
    - If no, where are the various points?
    - Is there any one-pass cooling water co-mingling with regulated wastewater at the sample point(s)?

7. Is there any batch dumping of: caustic tank (boil-out tank)?* Test tank?*
    - What frequency?
    - What volume?

8. Are repairs done over the test tank? Explain...

9. Is there any pretreatment of the waste water?
    - If yes, describe the various aspects in place and where located?
    - If no, is any required?

10. How is the routine disposal of hazardous waste handled - i.e., from the boil out tank? The sludge from the test tank?
    - Who is the waste hauler?
    - What is the average volume of material disposed of?
    - Where are the records kept?

11. Is there adequate spill containment in the process area(s)?
    - If yes, describe.
    - If no, describe.
    - How is the sewer otherwise protected from leaks or spills in the process area(s) and throughout the facility? Describe in detail.

* The bottom sludge must be handled as hazardous material - supernatant water can be discharged to sewer if pH and dissolved metal concentrations are within local sewer agency limits.

A-23

∆ **Steam Cleaning/High Pressure-High Temperature Washing,** as an operation, is rarely done in isolation but is usually an adjunct to some other industrial process; be that a machine shop, a car wash, a new car dealership, or other activity. If there is a degreasing job that will respond to high pressure or high temperature, then one or the other of these methods is probably the most reasonable choice. The use of either of these methods demands two things:

1. An isolated work pad which will contain wastes from the material being cleaned plus restrict rain water intrusion and

2. Some type of suitable grease interceptor installed in or adjacent to the cleaning pad.

A grease interceptor for this type of application usually does not have to be very large as the throughput of liquid is minimal. Nonetheless, the accumulation of greasy sludge can become excessive if the interceptor is not cleaned regularly. For this reason, the sumps or clarifiers should be easy to access for the purposes of routine inspection and maintenance.

INSPECTOR'S QUESTIONS—STEAM CLEANING/HIGH PRESSURE-HIGH TEMPERATURE WASHING

1. What type of material is being cleaned? Describe...
    - How is it being cleaned?
    - Any chemicals used?

2. Is there a grease interceptor that receives the cleaning wastewater?
    - If yes, what is the size of the interceptor?
    - Where is it located?
    - Is an elbow present at the discharge side?
    - If no, explain why absent?
    - Is this discharge properly routed to sewer? Explain...

3. How frequently is the interceptor cleaned?

4. How is the sludge disposed of?

5. Has any laboratory testing been done to establish the hazardous or non-hazardous condition of the clarifier sludge?
    -Describe and enclose copies of any lab reports if available...

6. How is the steam for the steam cleaner generated?
    - Boiler? (If yes, follow up with standard boiler questions)
    - Other?

7. How is the hot water for the High Temperature-High Pressure washer generated?
    - Boiler? (If yes, follow up with standard boiler questions)
    - Other?

8. Is there adequate spill containment in the process area(s)?
    - If yes, describe.
    - If no, describe.

- How is the sewer otherwise protected from leaks or spills in the process area(s) and throughout the facility? Describe in detail.

❏ Car Washes

Car washes are found as two major types:

- Full service or automatic car washes

- Coin-operated or self-service car washes

Under some protocols, only the first type, full service or automatic car washes, are eligible for industrial waste discharge permit. This selection may vary from city to city and is subject to local regulations. However, irrespective of local regulation, EPA requires all car washes over 25,000 gallons per day of discharge to be fully permitted.

Full service car washes generally operate within a discharge volume range of a few thousand gallons per day to more than 25,000 gallons per day in very large or very busy sites. The wide discharge range is due to two main factors:

- Volume of business

- Presence or absence of water reclamation systems

The effluent profiles from all car washes is usually rather benign but can show an occasional grease and oil or sludge buildup. The pollutants of concern at all types of car washes are mud and silt, from the clarifier bottoms, and the top grease and oil layer. These conditions can easily be monitored by inspection and/or sampling. Even emulsified oil problems can be detected by a careful inspection of the clarifier.

INSPECTOR'S QUESTIONS—CAR WASHES

1. What type of car wash is operated?
    - Full service?
    - Automatic?
    - Coin-operated?
    - Self-serve?

2. How many cars are washed per day?

3. Is there any water reuse or recycling at this facility?
    - Explain...

4. Is there any steam cleaning or high temperature-high pressure washing done at this site?
    - Explain...

5. Is there any rag washing done at this site?
    - Explain...

6. Are there any other operations done which use water and/or discharge to the sewer?
    - Explain...

7. Do all process streams flow to a common interceptor point?
    - If yes, where is that point?
    - If no, where do they discharge?

8. Is there any pretreatment of wastewater?
    - If yes, describe the various aspects in place and where located...

9. How frequently is the clarifier cleaned out?
    - Who is the hauler?
    - Where does the material go?

- Where are the records kept?

10. Has any laboratory testing been done to establish the hazardous or non-hazardous condition of the clarifier sludge?
    - Describe and enclose copies of any lab reports if available...

11. Is there any de-ionized water?
    - If yes, is it on a service? Or do you maintain the equipment?
    - What, if any, pretreatment is there for the systems?...Describe?
    - Is water kept in a holding tank and drawn off as needed?
    - Or is it produced on an as-needed basis?
    - Is the water metered in? Out?

12. What solvents are used at this facility?
    - Where are they used? What unit operations?
    - What is the method of application? Dip tank, rag applied, or other?
    - Is there any discharge of solvents to sewer in any form? From the unit operation(s)? Final triple rinsing of containers prior to disposal? Explain...
    - If no solvents are discharged to sewer, explain what procedures are in place to prevent discharge. Solvent Management Plan (SMP)?
    - When was SMP submitted? Explain...

13. Is there adequate spill containment in the processing area(s)?
    - If yes, describe.
    - If no, describe.
    - How is the sewer otherwise protected from leaks or spills in the process area(s) and throughout the facility? Describe in detail.

❑ Paint Shops/Body Shops/Detail Shops

Of all the permitted types of automotive dischargers, this group is by far the most inconsequential as far as being a <u>sewer</u> problem. This group of industries rarely ever needs a permit. They are usually noticed because they are discharging illegally to a street, alley, or in some other way causing a discharge that brings them to public attention. It is usually via a citizen complaint that they come to the inspector's attention.

There are no special requirements for these sites when discharging to sewer - the dilemma is getting them legally connected to sewer. Many times these businesses are located under tarpaulins, in sheds tacked onto legitimate structures or they are operating in vacant lots or parking structures. The legitimate business or building whose space they share, and whose water they use, rarely wants to become part of the "tenant's" permit process (which is usually unnecessary to begin with).

A better course of action in these cases is: (1) if the discharge does not leave private property, ignore the entire situation (it is a landlord/tenant problem at this point), or (2) if the wastewater discharge flows onto public property, it is properly a nuisance complaint. The perpetrator of the discharge has a choice of (a) keeping the water on his property, (b) halting water operations, or (c) legally connecting to sewer and obtaining a discharge permit if the conditions are such that one is mandated. Failure to comply with one of the options above can result in increased compliance actions against the discharger. These include possible referrals to Regional Water Quality, Fish and Game, County Hazardous Materials Management Unit, or County Environmental Health.

It is up to the inspector to determine which protocol or procedure will be implemented and to initiate the appropriate referral(s).

Section B

Food Processing—Including Fruits and Vegetables, Seafood, Slaughterhouses, Meatpacking and Dairy Products, and Bottled Beverages

Overview

Food processors, in general, share some common features which make the formation of a generic group for industrial waste discharge evaluation quite plausible. Chiefly, they all share the trait of producing no toxic waste water per se, but rather their waste water contains the more conventional pollutants such as grease and oil and heavy loads of suspended solids. Usually, the waste water has a high BOD as well. The pH is generally on the caustic side because of the many detergents and caustic cleaning compounds routinely used but can be acidic in some instances. An additional similarity is the across-the-board necessity for some type of steam production system, or other method for sterilization and cleaning procedures. Also needed is some type of cooling or refrigeration system to keep the processed foodstuffs from perishing. Lastly, they almost always will require some type of clarifier, settling vat or other device to remove the bulk of the solids from their effluent. Only rarely can a food processing site survive without some type of solids removal plan. The exception would be a very small operation or in a facility where only beverages are produced. Also, the pH of the discharge effluent can have wide swings from below 5.0, when acid disinfecting or cleaning solutions are being used to above 11.0 when strong caustic cleaning solutions and detergents are utilized. Without equalization tanks or systems in some cases it may be necessary to install pH control systems, either in lieu of or in addition to equalization tanks, when flows are large and/or swings are very wide.

INSPECTOR'S QUESTIONS—FOOD PROCESSING

1. What type of food processing is done at this site?
    - Fruits and vegetables - canning/preserving?
    - Seafood processing - fresh/preserved?
    - Slaughterhouse - beef/chickens/other?
    - Meat packing - beef/chicken/other?
    - Dairy products - milk/ice cream/yogurt/butter/etc.?
    - Beverages - soft drinks only?

   Describe each in detail including if canning, cooking, preserving, filleting, etc. is done and what operations are involved.

2. What types of washing operations are done?
    - Of the foodstuffs?
    - Of the process equipment?
    - Of the work area?
    - Other?

3. What types of detergents are used? Disinfectants? Describe for each type of washing done. Describe for each type of disinfecting done.

4. Is there routine washdown of the work area(s)?
    - Where is the discharge point?
    - Are there traps in the floor drains?
    - How often are the traps cleaned? How done? Explain...
    - Is an oil and grease or other type of interceptor present?

5. Are any types of processing brines used?
    - Are these brines sewered?
    - If yes, where is the discharge point?
    - What is the volume discharged?

6. Are larger remains of processed foodstuffs ground up and sewered?
    - Any attempt to restrict solids to sewer?

- Are the larger remains used as by-products?
- Explain how these materials are handled...

7. Are there any floor drains in the process area?
    - Does their flow pass through a common interceptor point?
    - If no, where does it discharge?
    - Is an interceptor or type of pretreatment chamber present?

8. Are there any cooling towers?
    - How many are there?
    - What is the tonnage of each?
    - What is the volume (gallons per day) discharged to sewer as bleed
    - where are the sample points?
    - What are the evaporative losses?
    - Any chromates used in the water treatment? Describe...

9. Any rectifiers, compressors or similar equipment in use?
    - If yes, do any of them require cooling water?
    - Any one-pass cooling water?
    - If yes, volume (gallons per day)?
    - Where is the discharge point?

10. Any requirements for steam generation or other high temperature water use?
    -If yes, is make-up water metered?
    -Volume (gallons per day)?
    -Is there blowdown to sewer? ...Frequency? ...Volume (gallons per day)?
    -What are the steam losses (gallons per day)?

11. Is any steam cleaning or high temperature/high pressure washing done? Describe and include sample point if applicable..

12. Is there any de-ionized water?
    -If yes, is it on a service? Who maintains the equipment?
    -What, if any, pretreatment do you have for the DI system?...Describe?
    -Is water kept in a holding tank and drawn off as needed?

-Or is it produced on an as-needed basis?
- Is the water metered in? Out?

13. Any other type of water treatment? Soft water, etc.? Describe...

14. Do all process streams flow to a common interceptor point?
    - If yes, where is that point?
    - If no, where are the various points?

15. Is there any pretreatment of waste water?
    - If yes, describe the various aspects in place and where located.

16. Is there any water reuse in the plant?
    - If yes, where and what is it?

17. Is there an ice machine at this site?
    - What volume of water is used per day for making ice?

18. Is any water incorporated into the product?
    - Where and what is the average gallons per day volume?
    - What is the rate of return to sewer when incorporation into product is considered?

19. Is there any other specialized water treatment done for incorporation into or with the product produced? Explain...
    - Does this treated water by-product discharge to sewer?
    - If yes, where...
    - How many gallons per day?
    - Are there any especially high loadings of suspended solids produced by this treatment process?

20. Is there adequate spill containment in the process area(s)?
    - If yes, describe.
    - If no, describe.
    - How is the sewer otherwise protected from leaks or spills in the process area(s) and throughout the facility? Describe in detail.

Section C

Hospitals

Introduction

Even though a wide variety of toxic, hazardous and prohibited materials are used by (and discharged by) hospitals, the overall quantity disposed of to sewer is relatively low and serious pollution problems are usually non-existent. Nonetheless, a poorly managed facility has the potential to cause serious problems for the sewer department. These problems could run the gamut from being merely troublesome, if a grease blockage was the concern, to constituting a major issue if concentrated acids, flammables or explosives were involved.

Thankfully, this latter condition is not the norm. In this section, various aspects of hospital operations which produce industrial waste water will be examined in some detail. They are as follows:

- Laboratory - Clinical, Pathology, Bacteriology

- Morgue

- X-ray Department and Nuclear Medicine

- Physical Therapy

- Laundry/Central Services

- Diet Kitchen/Cafeteria

- Boilers

- Soft Water Production

- Reverse Osmosis and De-Ionized Water Production

- Cooling Water Systems/Chillers

- ❑ Vehicle Washing/Transportation
- ❑ Photo/Printing/Graphic Arts
- ❑ Hazardous Waste Disposal/Storage

❑ Laboratory: Clinical - Histology - Cytology - Bacteriology

Even in the smallest hospital it is usual to find at least a clinical laboratory where traditional blood, urine and other body fluid analyses are done. In the larger, newer or more sophisticated institutions, all of these same analytical procedures will also be present but with two major differences: (1) Many more individual types of testing procedures will be performed which provide information to assist in making differential diagnoses, and (2) the analytical equipment in use will undoubtedly be computerized, more sophisticated and use less water and chemistry than a smaller or older counterpart. Also, the larger facility is more apt to have associated with it the ancillary laboratory sections of bacteriology/microbiology, histology, cytology and perhaps even toxicology and virology labs, vivarium, or other specialized subgroups related to clinical medicine.

Clinical Laboratories usually account for a very low volume of any significant industrial wastes. Their usual effluent is diluted solutions of body fluids in combination with chemical reagents ranging from simple isotonic saline to low volume, low concentration cyanide solutions. However, it is not uncommon for modern day laboratories to either isolate all of their industrial flow and pass it through an interceptor where acid neutralization can take place or to individually plumb all lab sinks to acid neutralization units. The technology of clinical medicine has advanced to such a degree that analysis which 10-20 years ago required a few milliliters or so of reactant to derive data now require only micro-amounts. For this reason alone, nothing of much significance leaves the clinical lab via the sewer. However, in those institutions where laboratory glassware is still washed (all items are not disposable), the washing operation needs close examination. It is from this area that the most problematic discharges can originate if chromic or other acid cleaning of any glassware is done. It is common practice to soak heavily soiled pipettes or other glassware in this type of acid solution. Many times the employees washing the glassware are not as alert to the adverse effects of an acid discharge to drain as they should be and the glassware washing area, unlike the laboratory proper, is not usually plumbed with acid neutralization units on the discharge.

The **histology laboratory** is where thin tissue sections are examined either via frozen section performed during a surgical procedure for the advance diagnosis of cancer or during routine examination of post-mortem (autopsy) tissue and body parts for determining cause of death.

In many instances the **cytology laboratory** is in close association with the histology lab because of the related nature of their work. The cytology laboratory processes and examines PAP smears looking for the presence of cancerous or pre-cancerous cells.

The effluent produced by these two closely related disciplines is also very similar. However, the cytology laboratory output is usually smaller in scale. The chemicals of concern here are xylene, formaldehyde and, to a lesser extent, acetone, acid-alcohols and organic stains used in the processing of the gross tissue sections and diagnostic slides. Most laboratories no longer discharge xylene to the sewer. The case has been made for (1) collecting back all of the used xylene for hazardous waste disposal or recycling or (2) substituting the xylene with an analogous product which can usually be sewered in small amounts. (E.g., Hemo-D.)

Also, low volume solutions of 4-10% formalin are sometimes allowed to be discharged to the sewer depending on the local sewer ordinance, but it is more environmentally sound to handle it as hazardous material.

Bacteriology/microbiology usually produces no industrial wastewater, only biohazardous wastes which are handled via their own in-house procedures for such wastes. This includes double-bagging in red biohazard bags which must always be autoclaved prior to disposal. The autoclave can use single-pass or other cooling water and this is usually the only significant water use from this section.

The **toxicology lab** deserves special mention because they are the only discipline left which may still do the "wet chemistry" type of analysis. What this means is that larger volumes of organic solvents could find their way into the sewer from these laboratories. Also, many times the toxicology lab may have an animal study program associated with it and then the vivarium has to be included in the inspection. The ongoing maintenance of a vivarium entails daily cleaning and disinfection procedures which can generate a discharge to sewer and may also require some type of solids removal equipment.

❑ **Morgue**

Generally speaking the only discharge from the morgue is a result of the routine cleanup after post-mortem examinations are performed. Standard cleaning and disinfecting chemicals would be used. The only pollutant of concern leaving this area is most apt to be a biohazard. This biohazard would take the form of a disease pathogen entrained in or associated with a discharged body fluid such as blood.

❏ X-ray Department and Nuclear Medicine

The standard X-ray film is nothing more than a variable size black and white negative and it is processed in a similar fashion. All hospitals currently use automatic processors for their medical X-ray operations. The automated processor(s) need to be evaluated individually and collectively for permitting purposes. All of the same parameters which would apply to photo processing apply here including the obligation to provide silver recovery where appropriate. (See Black and White Film Processing and the associated diagram for additional details.)

The **Nuclear Medicine** lab is closely related to the X-ray department and is dedicated to a specialty study area of radiology. Radioactive "cocktails" or "tagged" solutions are administered to patients in order to evaluate the function and/or performance of particular body organs. As the organ or tissue takes up the radioactively tagged material, functioning parameters can be monitored and evaluated using various nuclear medical techniques.

Three of the most common radioactive isotopes administered clinically are I^{123} (Iodine 123), Gallium and Technesium 99. The presence of this type of radiation in wastewater is monitored by beta radiations analysis. The presence of alpha radiation in a routine medical facility's wastewater is not generally a concern since the radiological sources are usually beta (or gamma emitters) - not alpha emitters.

❑ **Physical Therapy**

No pollutants of significance are discharged from this section. However, if a hospital has a sizeable physiotherapy unit, water consumption on its own could be a factor which needs critical evaluation. Also, traditional cleaning and disinfecting chemicals are routinely used to wash out the hydro baths, whirlpools, etc. This is of particular concern when severely burned patients are using the spas or therapy pools.

❏ Laundry/Central Services

The trend has been for hospitals to eliminate in-house **laundry services** and hire this service out to large commercial firms. More than likely economic factors have entered into this decision as the prices for labor, electricity and water have risen and outdated laundry equipment required replacement. Nonetheless, it is not impossible to find instances where hospital linens are washed on-site even if it is only some relatively low volume specialty linen.

(If a laundry service is included follow the guidelines presented in the Commercial Laundry section to fully evaluate their industrial wastewater discharge.)

The other related service usually present at all hospitals is **Central Service** or **Central Supply**. Their primary function is to provide sterilized materials to the hospital. This type of material is most often used in the surgical arena but there is also application in other areas of patient care where completely sterile material must be used and sterile environments must be maintained. Two of the most common sterilization procedures used are: (1) Steam autoclaving and (2) ethylene oxide gas sterilization.

If an autoclave is used there may be cooling water associated with it and this should be evaluated. If ethylene oxide gas is used, normally there is no extraordinary concern for the effluent produced even though the gas itself is formidably toxic. Extreme care must be taken in the use of this gas and any possible exposure to handlers prevented.

❑ Diet Kitchen/Cafeteria

The two principle considerations in the inspection of the food services operations are:

- Water Use

- Grease Handling and/or Interceptors and Grease Traps

"Full service" kitchens in large hospitals can be significant contributors to the total wastewater flow. It is generally accepted that 8 gallons of water per meal is a good estimate for producing effluent flow data. This would include the flow from automatic dish washers and pot scrubbers. The other important consideration is how the grease traps and/or clarifiers are maintained. If none are identified, the determination of grease control requirements must be done. In lieu of grease removal equipment, some sites merely incorporate drain cleaning chemicals into their routine maintenance schedules. Most times these drain cleaning chemicals are either strong caustics, strong acids or chlorinated hydrocarbons, or some combination. None of these chemicals are desirable contributions to the sewer system and they are prohibited by some sewer ordinances because of the problems that can be caused by their introduction into the sewer lines.

❑ **Boilers**

* See Boiler Section

❑ **Soft Water Production**

* See Boiler Section

❑ **Reverse Osmosis/De-ionized Water Production**

* See Treatment Technology Section

❑ **Cooling Water Systems**

* See Cooling Water Section

❑ **Vehicle Washing/Transportation**

* See Automotive Section; Car Washes, Steam Cleaning, etc.

❑ **Photo/Printing/Graphic Arts**

* See Photo Section
* See Printing Section

❑ Hazardous Waste Disposal/Storage

There is **normally** no discharge from the storage area and there never should be. The only reasons for including it here are:

- The inspector should determine that there is a properly maintained area for storage of waste hazardous materials primarily to insure that these materials are not being discharged to sewer out of ignorance or laziness.

- The inspector should examine the storage area and make sure that no floor drains, sumps, or other sewer access is present.

- If the storage area is outside in a fenced area and not completely bermed or otherwise closed off, the inspector should evaluate the area for possible spills to the ground and/or storm drain system and make the necessary referrals to other agencies if corrections are not made.

If there is any disposal of "specialty" wastes, as can be the case with certain decayed radioactive materials, or if some other centralized waste treatment facility exists on-site, the inspector should also include this in the site inspection and evaluation to be sure that all procedures are legal and proper and correctly allocated on the discharge permit.

INSPECTOR'S QUESTIONS—HOSPITALS

1. Describe the Hospital as to type (full-service medical, surgical, skilled nursing facility, etc.)...

2. Identify the services offered and evaluate each for discharge...
    - X-ray - Nuclear medicine?
    - Laboratory? (Itemize all...)
    - Physiotherapy?
    - Laundry?
    - Central Supply?
    - Food Services?

3. Is there any reverse osmosis or de-ionized water production on site?
    - If yes, is it on a service? Who maintains the equipment?
    - What, if any, pretreatment do you have for the de-ionization system?...Describe?
    - What is the reverse osmosis reject ratio? (I.E., what percent of the reverse osmosis goes back to the sewer as reject and what percent is used as process water?)
    - Is water kept in a holding tank and drawn off as needed?
    - Or do you produce on an as-needed basis?
    - Is the water metered in? Out?

4. Is there any other type of water treatment in use at this site? Softwater production, etc.? Describe...

5. Any rectifiers, compressors or similar equipment in use?
    - If yes, do any of them require cooling water?
    - What type of cooling systems are present? Describe...
    - Any one-pass cooling water?
    - If yes, volume (gallons per day)?
    - Where is the discharge point?
    - If water is not used to cool equipment describe cooling practices.

6. Is there any water reuse in the plant?
    - If yes, where and what is it?

7. Any requirements for steam generation or use?
    - If yes, is make-up water metered?
    - Volume (gallons per day)?
    - Is condensate returned to system?
    - Is there blowdown to sewer? ...Frequency? ...Volume (gallons per day)?
    - What are the steam losses (gallons per day)?

8. Are there any requirements for steam generation or other high temperature water use?
    - If yes, is make-up water metered?
    - Volume (gallons per day)?
    - Is there blowdown to sewer? Frequency? Volume (gallons per day)?
    - What are the steam losses (gallons per day)?

9. Is any steam cleaning or high temperature/high pressure washing done?
    - Describe and include sample point.

10. Is there any vehicle washing at this facility?
    - If yes, describe and include description of pretreatment equipment present and sampling location.

11. Is there any routine washdown of the work area(s)?
    - Where is the discharge point?

12. Are there any floor drains in any process area(s)?
    - Does their flow pass through a common interceptor point?
    - If no, where does it discharge?

13. Do all process streams flow to a common interceptor point?
    - If yes, where is that point?
    - If no, where are the various points?

14. Is there any pretreatment of waste water?

- If yes, describe the various aspects in place and where located.

15. How is routine disposal of hazardous wastes handled?
    - Who is the waste hauler?
    - What is the average volume of material disposed to landfill, recycled, etc.?
    - Where are the records kept? (Can ask to see manifests.)

16. Is there adequate spill containment in the process area(s)?
    - If yes, describe.
    - If no, describe.
    - How is the sewer otherwise protected from leaks or spills in the process area(s) and throughout the facility? Describe in detail.

Section D

LABORATORIES

Introduction

There are many different types of laboratories which use and discharge small amounts of toxic and prohibited materials. Laboratories can be located in medical offices, clinics, private businesses, or hospitals. A decision to inspect and permit laboratories is made based on what the local sewer ordinance and/or current protocol requires. Some common types of laboratories which could require industrial waste evaluation are the following:

- Medical/Clinical Laboratories*
- Research Laboratories
- Toxicology Laboratories
- Chemistry Laboratories
- Pathology/Histology Laboratories*
- Commercial Analytical Laboratories
- X-ray Laboratories*

Also in this section the following laboratory-related topics will be covered:

- Water Uses
- Constituents of Concern
- Inspector's Questions

* See Hospital Section for additional detail on these labs.

❑ **Water Uses**

With the advent of new water conservative equipment, most laboratories have seen their water uses significantly reduced. The two primary uses of water in most laboratories today are for:

- Glassware and equipment washing

- Single-pass cooling water needed for various pieces of analytical equipment or chemical procedure

To a much lesser degree there is also some area washdown and routine rinsing of used or dirty containers, beakers, and other miscellaneous pieces of laboratory glassware prior to washing.

❑ **Constituents of Concern**

Regardless of laboratory type, usually the most important constituents of concern as far as sewer discharge is concerned are:

- Discharge of strong acids

- Discharge of potentially combustible, flammable, and/or toxic solvents such as alcohol, benzene, acetone, xylene, or ether

The first case problem can easily be eliminated by the routine installation of marble chip neutralization units at every lab sink in a facility or a large interceptor with the ability to equalize or chemically treat acid wastes. This coverage should include the glassware washing room where occasional significant discharges, especially of chromic acid, can and do occur. Chromic or other acid is traditionally used to acid clean pipettes and other pieces of laboratory glassware requiring scrupulous cleanliness.

The mitigation of flammable or combustible solvent discharges to the sewer is better addressed by increased efforts at recycling product, recovery or product substitution. Fortunately, most laboratories do not use or discharge excessive amounts of flammable or combustible materials to the sewer.

Any other hazardous toxic or prohibited materials which are used in the laboratory, or produced as by-products, are not usually a problem in this day and age. However, the inspector needs to be ever alert to that "once in a blue moon" discovery of chemicals such as benzene or cyanide, or an aging can of ethyl ether or bottle of picric acid found during the inspection tour which could be handled dangerously or inappropriately.

INSPECTOR'S QUESTIONS—LABORATORIES

1. What type of laboratory operations are conducted?
    - Medical?
    - Research?
    - Analytical? Describe...

2. Which areas of the laboratory have water and/or chemical discharges to the sewer?
    - Identify those areas and describe the unit operations which generate waste chemical or wastewater discharge to the sewer...

3. Is any wet chemistry performed?
    - If yes, describe...

4. Which areas use solvents in their procedures?
    - Identify them and describe how the solvents are used...
    - Identify which solvents are discharged to sewer...
    - Identify which solvents are collected for hazardous waste disposal or recycling (on-site distillation) or other method of disposal...

5. Is routine glassware washed or disposable?
    - Identify which material is disposable...
    - identify which glassware is washed...

6. Is any glassware "acid-washed"?
    - Is chromic acid used?
    - Any other acid used in final rinse? Describe...
    - Is it discharged to sewer?
    - If yes, is there any pH neutralization associated with this discharge?

7. How are spent or contaminated chemicals handled?
    - If hauled away, who is the hauler? (Can check manifests)

8. Does any of the laboratory equipment require cooling water?
    - Any rectifiers, compressors, or similar equipment in use?
    - Any one-pass cooling water?
    - If yes, volume in gallons per day?
    - Where is the discharge point?
    - If not one-pass, is there a cooling tower in use?

9. Is there any X-ray film processing done?
    - How many processors?
    - How many hours/day in operation?
    - What is the daily average flow from each processor in gallons per day?
    - Where is the discharge point(s)?
    - Are there other waste streams discharging into that same point?
    - Is there any silver recovery used?
    - If yes, describe the type...
    - If no, explain why no recovery is in place...

10. Is there any routine washdown of the work area(s)?
    - Where is the discharge point?

11. Are there any floor drains in the process area?
    - Does their flow pass through a common interceptor point?
    - If no, where does it discharge?

12. Is there any reverse osmosis or de-ionized water?
    - If yes, is it on a service? Who maintains the equipment?
    - What, if any, pretreatment is there for the systems?...Describe?
    - What is the reverse osmosis reject ratio? (I.E., what percent of the reverse osmosis goes back to the sewer as reject and what percent is used as process water?)
    - Is water kept in a holding tank and drawn off as needed?
    - Or do you produce on an as-needed basis?
    - Is the water metered in? Out?

13. Do all process streams flow to a common interceptor point?
    - If yes, where is that point?

- If no, where are the various points?
- Is any one-pass or brine reject water co-mingling with regulated wastewater at the sample point(s)?

14. Is there any pretreatment of waste water?
    - If yes, describe the various aspects in place and where located.

15. Is there any water reuse in the facility?
    - If yes, where, and what is it?

16. Is there adequate spill containment in the process area(s)?
    - If yes, describe.
    - If no, describe.
    - How is the sewer otherwise protected from leaks or spills in the process area(s) and throughout the facility? Describe in detail.

Section E

LAUNDRIES - INDUSTRIAL AND COMMERCIAL

Introduction

In general, laundries covered in this section will be of two main types: (1) Industrial laundries which process heavily soiled items like rags, rugs and uniforms from "dirty" occupations such as mechanics and laborers, and (2) commercial laundries which handle lightly soiled articles similar to bed and table linens and uniforms from "clean" occupations such as automotive sales personnel and fleet drivers. Carpet and upholstery will be considered the same as commercial laundries, as will diaper services. Coin-operated laundries will not be discussed. For discussion purposes, an industrial laundry will be considered to process <u>at least 20%</u> of the industrial type of article (e.g., rags, mop heads, rugs, etc.), otherwise the classification will be considered to be commercial. The following areas are covered in this section:

- Industrial Laundries

- Commercial Laundries

- Water Uses

- Rates of Return

- Constituents of Concern

- Pretreatment of Waste Water

- Inspector's Questions

❑ **Industrial Laundry**

Because of the type of garments cleaned in an industrial laundry, the facility's waste water is typically contaminated with high levels of grease and oil, heavy metals and a variety of organic solvents. In other words, their effluent quality is a direct reflection of the type of work being processed at any given time. This profile can change hour to hour and day to day depending on the work schedule and/or work flow. However, it is usually the shop rag component of their business which causes most of their problems with effluent quality.

The industry as a whole has been slow to come into compliance with local discharge limits (NOTE - see Figure E-1) across the nation. The industry is quick to point out that tough competition, economic constraints, and a lack of treatment systems and technology to meet their specific needs have all influenced the EPA's decision not to regulate them as a categorical industry. However, it does seem likely that launderers will come to be federally regulated, probably within the next decade.

The backlash of EPA's decision not to regulate launderers may have caused more problems for the industry as a group than it has eliminated. Given present conditions, the laundries have to cope with widely varying standards across the country. Dependent on the local limits imposed, a nationally franchised firm has little incentive to develop treatment systems that will be guaranteed to meet compliance limits in different states, counties or even cities. Having an even more negative impact on their business is the local sewer agency who imposes unrealistic or unattainable discharge limits on the laundry permittee, perhaps out of a sense of needing to impose numerical limits. The permittee has the obligation, by local ordinance, to meet these limits or be out of compliance with their discharge permit. The only other option is to convince the local authorities to change the sewer discharge limits to something more reasonable and/or attainable with current pretreatment strategies. None of these are easy choices.

ESTIMATED PERCENT OF LAUNDRIES HAVING CONTROL TECHNOLOGY

(From *Industrial Launderer*)

Control Technology	% installed
Bar Screens	2.7
Lint Screens	70.0
Catch Basins	72.0
Heat Reclaimers	70.0
Oil Skimmers	15.0
Equalization Tanks	1.2
pH Adjustment	4.1
Physical/Chemical Systems	1.3
* Other	8.2

* Other includes filtration, separators, oil hold-back devices, and misc. operations

Figure E-1

❑ Commercial Laundry

The typical effluent quality from a strictly commercial laundry is more static than is its industrial counterpart. The usual variations are found in temperature, solids, load and pH level. All of these changes in parameters can be traced back to what was happening in the plant at the time of measurement; if it was washing, rinsing or other activity.

The typical garments found in commercial laundries include hotel, motel, hospital and other institutional linens, gymnasium towels, uniforms and lab coats from commercial and medical professions and similar lightly soiled garments. Though seemingly a contradiction, diaper services are also included in this category.

❏ Water Uses

The U.S. Public Health Service estimates that laundries, as a group, use four gallons of water per pound (dry weight) of material processed. In 1979, the EPA conducted studies which showed an average rate of 4.8 gal/lb. and laundry industry figures show 3.7 gal/lb.

The breakdown for the figures produced by the laundries is as follows:

Total Water Used	3.7 gal/lb
Washers	2.8 gal/lb
Cooling equipment	.5 gal/lb
Boiler make-up	.3 gal/lb
Sanitary	.07 gal/lb

Also see Figure E-2.

• Laundering

It can be safely assumed that about 90% of the water use at any given plant goes exclusively to the laundering process, with the remaining 10% lost to evaporation. As shown in the various studies conducted, there is a variation in the average water consumption per weight of material processed. This variation is understandable considering that different plants will be laundering different types of garments with varying degrees of soil. Older plants will have less efficient water use and newer equipment will be more water conservative. All of these factors can combine to produce a somewhat different gallons/pound average. However, overall approximately 4-5 gallons/lb is typical and 10% losses are reasonable.

TYPICAL COMMERCIAL/INDUSTRIAL WATER DISTRIBUTION DIAGRAM

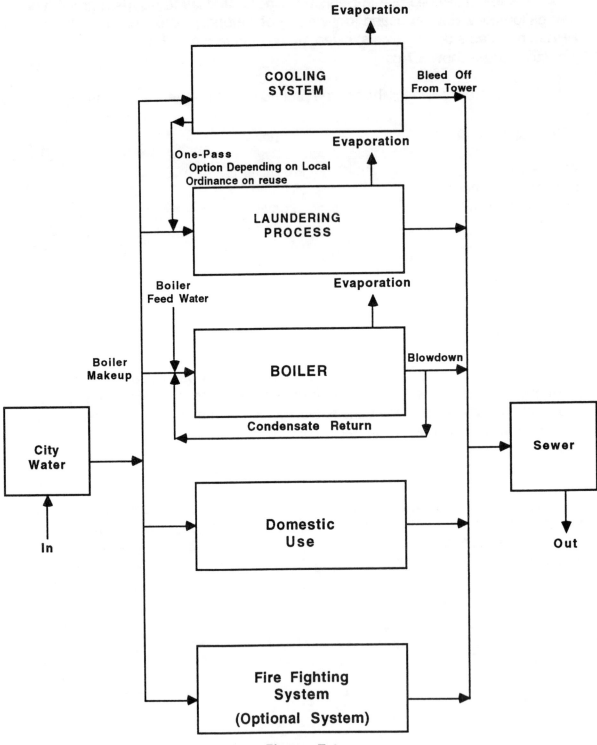

Figure E-2

- **Boilers**

    An integral part of any laundry operation is the boiler. The boiler is needed to produce the heat and/or live steam needed in the plant.

    From an inspector's perspective the boiler system is important for the following reasons:

    Δ Boiler water use is a significant part of calculating the water audit for the plant. Water is consumed and of that a certain percentage is lost to evaporation and another is discharged to sewer as blowdown.

    Δ Boiler feedwater usually requires some type of treatment prior to being used in the system. The specifics of this treatment can add additional wastewater streams into the total industrial flow. The usual water treatment is sodium zeolite softening, but other procedures can also be used.

- **Cooling Systems**

    Many times there is a requirement for cooling water at a laundry as well as heated water. The most common application is for cooling the dry cleaning still condensers. Conventionally, dry cleaning fluid (usually perchlorethylene) is recycled on-site and in the process it is necessary to recondense the recycled fluid prior to reuse in the dry cleaning drum. The cooling water for the condensers could be (1) one-pass which is then re-routed to the laundering process for use there, or (2) on a looped system supported by a cooling tower.

- **Fire Fighting Systems**

    While these systems use no water per se, there may be a reservoir on-site for this specific purpose. Also, it is not uncommon that the sprinkler systems or other fire fighting devices require periodic testing and some minor water use can be allocated to that.

❏ **Rates of Return**

Commercial and industrial laundries are one of the industry groups which consistently shows an across-the-board evaporative loss to be present. This loss can vary somewhat due to individual circumstances, but a good average is 9-10%.

Nationwide, the data collected by both industry and government indicates 10% average for evaporative losses.

How the sewer agency bills their customers for rates of return to the sewer will directly impact the industrial/commercial launderer. If no attempt is made to identify the losses, and billing practices do not take evaporative losses into consideration, the laundry could be overcharged for sewer fees. If, on the other hand, billing practices assume even lower than normal rates of return for commercial and industrial customers, the sewer agency could lose revenue. It is very important that accurate billing codes and practices are established and enforced. Only in this way are both the customers and the sewer agency assured that reasonable fees for services are implemented. In addition, the EPA has mandated that services received must be paid for in an equitable way. Categorizing laundries with respect to their rates of return to sewer is one way of making the fees for services equitable.

❑ Constituents of Concern

At both commercial and industrial laundries, pH can be a problem. This is especially true if there are no equalization pits or clarifiers present. Also, if a commercial laundry has a high volume of restaurant business their grease and oil level can approach or exceed local limits.

More commonly, it is the industrial laundry which has a consistent problem in meeting many of the locally imposed sewer limits. In addition to the pH and grease and oil problems, they can exceed both zinc and lead limits (and sometimes other metals as well). Also, organic constituents can and do show up in analytical profiles. Since the industry group is not federally regulated, it is sometimes difficult to evaluate what these figures mean as far as compliance is concerned - unless there is a local limit imposed which details in whole or in part which organic compounds are prohibited in the sewer and in what concentrations. It is more usual to have the situation whereby organics are limited, or prohibited, as they can cause combustion or flammability in the sewer system. More than likely this mode of thinking will change during the next decade as both sewer agencies and industries become more subjected to increased environmental regulations dealing with the release of organics into the atmosphere.

❏ **Pretreatment**

Depending on the type of garments washed, a laundry could have next to no pretreatment required to meet its permit limits or could be eligible for enormous outlays of capital costs for installing and maintaining sophisticated pretreatment systems.

At the very least, most laundries have some type of rudimentary lint trap, screening device or clarifier to remove the larger, heavier solids.* This is done to prevent excessive solids from entering the sewer and causing obstructions of flow. In addition, they commonly have some type of heat reclamation and maybe even a water recycling system or methodology. (Though there is usually an ordinance imposed upper limit on the effluent discharge temperature, (65° F - 150° F), heat reclamation systems are installed strictly for energy recovery, <u>not</u> as high temperature control devices.) A common heat reclamation technique involves preheating incoming water with outgoing water. Also, a typical water reuse procedure is to take one or more of the final rinse water discharges and recover and re-route them for use in the first wash cycle of the next load being processed.

The further addition of a large holding or settling tank and/or some type of equalization vessel provides a degree of retention and equalization for the waste water. This is a particularly valuable addition to their wastewater treatment. The variations in flow characteristics are so wide, virtually hour to hour, that any type of moderating influence is highly desirable.

Industrial laundry wastewater is often contaminated by high levels of grease and oil (it averages 600 to 1,000 mg/l), and of this amount, the free oil may only be 10% or so. The free oil is easily removed by skimming, either by manually dipping it off the water's surface inside the equalization tank, or by automatic removal via belt or rope skimmers. The free oil is a small problem. It is the emulsified oil which represents the major dilemma for industrial laundries to resolve. Also, it is this emulsified portion that usually has the remaining portion of heavy metals associated with it. (The other portion is typically found in with the free oil.) The oil and metals are bound up in the form of a colloidal suspension.

* Able to remove approximately 20% at best.

Chemical addition is currently the procedure of choice used to aggregate the colloidal material into particles which can then be removed more easily from the wastewater. The particles are physically removed using flotation, sedimentation, or filtration methods.

In a typical wastewater treatment system (see Figure E-3) chemicals are used throughout the treatment process and serve several distinct purposes. The initial addition of specific treatment chemicals breaks the water/oil emulsion. Caustic can then be added to aid in the formation of insoluble metal hydroxides and to start the coagulation process. Coagulation generally proceeds best at a pH of 11.2 - 11.5. The next step is to add the polymer as a flocculating aid. The treated effluent is passed through a dissolved air flotation unit (DAF) and the free grease and oil, suspended solids and metal hydroxides are removed. The sludge removal from the DAF is usually less than 10% solids and therefore a second solids enhancement step, the sludge wet well, is needed prior to the final dewatering and sludge cake production.

TYPICAL INDUSTRIAL LAUNDRY WASTEWATER TREATMENT SYSTEM

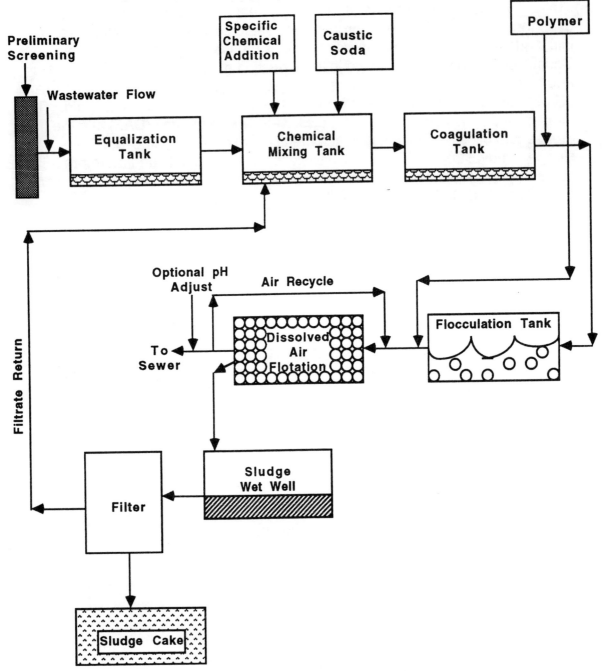

Figure E-3

INSPECTOR'S QUESTIONS—LAUNDRIES

1. What is the ratio of industrial product to commercial product within the washing process? (Typically only a very low percentage of industrial garments or materials is allowed if the site is to be classified as a commercial laundry.) Describe the product mix.

2. How many washers are in use?
    - What size? (Capacity in pounds)
    - How many pounds of laundry are washed per day?

3. Is there a steam tunnel in use?
    - Describe...

4. Is any dry cleaning done on-site?
    - If yes, is the dry cleaning solvent reclaimed on-site?
    - If yes, describe the reclamation procedure and include any cooling process and any wastewater that is involved with it...

5. Is any heat reclamation done on-site?
    - If yes, describe how and where done...

6. Does any process or piece of equipment require cooling water?
    - If yes, describe what and where done...
    - Is the cooling water from a closed-loop system (cooling tower) or from City water (single-pass)?
    - Describe...

7. Are there any cooling towers in use at this site?
    - If yes, how many?
    - What is the bleed frequency? Volume?
    - What is the make-up volume?
    - Is the make-up water metered?
    - Are any chromates used in the water treatment? If yes, describe

sample point.
- What chemicals are used?
- Where is the discharge point?

8. Are there any boilers in use at this site?
    - If yes, how many?
    - What is the type? (High pressure, low-pressure, fire tube or water tube, etc.)
    - What is the blowdown frequency? Volume?
    - What is the make-up volume? Is the make-up water metered?
    - Where is the discharge point?
    - What are the estimated losses?

9. Is there any softwater (sodium zeolite) produced on-site?
    - If yes, is it on a service or is the system regenerated in-house? Describe...
    - Is the soft water kept in a holding tank or produced on an as-needed basis?
    - Is this water metered in? Metered out:?
    - What is the volume of brine discharged to the sewer during the regeneration cycle?
    - How frequently is regeneration done?
    - Where is the discharge point?

10. Is there any reverse osmosis or de-ionized water production on-site?
    - If yes, is it on a service? Who maintains the equipment?
    - What, if any, pretreatment is there for the systems?...Describe?
    - What is the reverse osmosis reject ratio? (I.E., what percent of the reverse osmosis goes back to the sewer as reject and what percent is used as process water?)
    - Is water kept in a holding tank and drawn off as needed?
    - Or do you produce on an as-needed basis?
    - Do you meter this water in? Out?

11. Is there any routine washdown of the work area(s)?
    - Where is the discharge point?

12. Are there any floor drains in the process area?
    - Does their flow pass through a common interceptor point?
    - If no, where does it discharge?

13. Do all process streams flow to a common interceptor point?
    - If yes, where is that point?
    - If no, where are the various points?

14. Is there any water reuse in the plant?
    - If yes, where and what is it?

15. Is there any pretreatment of waste water?
    - If yes, describe the various aspects in place and where located.

16. How is the routine disposal of hazardous wastes handled?
    - Who is the waste hauler?
    - What is the average volume of material disposed to landfill, recycled, etc.?
    - Where are the records kept? (Can ask to see manifests.)

17. Is there adequate spill containment in the process area(s)?
    - If yes, describe.
    - If no, describe.
    - How is the sewer otherwise protected from leaks or spills in the process area(s) and throughout the facility? Describe in detail.

Section F

PAINT AND INK FORMULATION

Introduction

While there are some operations that are unique to every site, most paint and ink production plants have many activities that are common to all. Primary among them is equipment cleaning which alone can produce the bulk of the permittable wastewater effluent. (See Figure F-1.)

Equipment cleaning usually produces two distinct waste streams:

- **Spent solvent** from solvent rinsing operations when oil-based materials are produced

- **Watery paint wastes** from the high-pressure water and alkaline cleaning procedures used with latex products

When properly segregated, both latex water-based paint wastes and solvent wastes can be distilled on-site and a majority of the materials, including the water, reused. However, distillation residues, from solvent cleanup, are still required to be handled as hazardous materials. Also, by further segregating the solvents, the ability to recycle each solvent is improved. Better yet, using only one solvent produces a single waste stream that is even easier to handle.

The primary source of the watery/latex paint waste is the portable tank cleaning operation. The usual sequence is (1) manual cleaning of the tanks with spatulas or scrapers to remove any dried or clinging paint (the preferred procedure is to clean the tanks as soon after use as possible to minimize this step), and (2) rinsing with high pressure water accompanied by an alkaline cleaning solution.

The paint waste and wastewater produced by these two operations are routed to a clarifier where, over time, adequate separation occurs. The clear top water layer can be reused or sewered at this point. The current technology allows that the settled solids can be blended with additives (after some flocculation) to produce a beige-colored product which is sold as a general purpose paint. Thus, by reworking the solids residue, disposal is avoided

PAINT and/or INK PRODUCTION SCHEMATIC

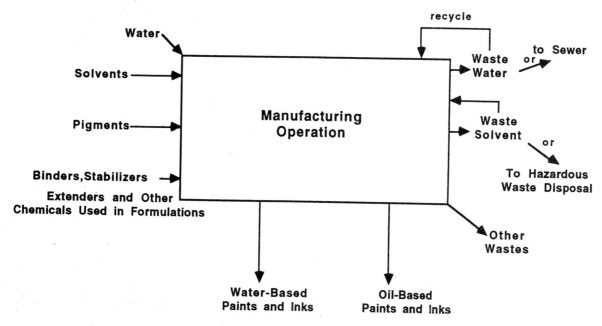

Figure F-1

altogether. This type of procedure is highly desirable both from an economic and an environmental perspective.

❑ **Constituents of Concern**

Normally, if a site is producing oil-based products only, no discharge to the sewer is expected. All of their cleanup will be done with solvents and all of their waste products will be recycled or disposed of as hazardous wastes. On the other hand, if water-based materials are being produced, certain heavy metals can be found in the wastewater. The most commonly occurring metals are titanium (white paint), chromium (yellow paint), lead (red paint), and mercury (a bacteriocide). Occasionally, copper can be found if blue ink is produced. It is best to ask for Material Safety Data Sheet (MSDS) on all materials where specific formulations are not known and sewerable effluent is being produced.

INSPECTOR'S QUESTIONS—PAINT AND INK FORMULATION

1. What types of paints or inks are produced?
    - Oil-based?
    - Water-based?

2. Are solvents used?
    - If yes, which ones are used?
    - How is the solvent kept out of the sewer? Explain...
    - Where are the solvents used?
    - How are they handled, disposed of, etc.?

3. Describe how oil-based equipment is cleaned...

4. If water-based materials are produced, what materials are used in their formulation?
    - What heavy metals from various pigments?
    - What heavy metals from chemical additives such as mercuric bacteriocides, etc.?

5. Describe how "water-based" equipment is cleaned...
    - Frequency?
    - What chemicals used?
    - Any special equipment?

6. Is there any discharge to the sewer via routine area washdown, spills, bad product, etc. Explain...

7. What are the individual volumes of the process tanks?

8. Are there any floor drains in the chemical storage area?
    - The processing area?
    - The cleanup area? Describe...

9. How are spent or contaminated processing tanks handled?
   - If hauled, who hauls it? (Can check manifests.)

10. Any rectifiers, compressors or similar equipment in use?
    - If yes, do any of them require cooling water?
    - Any one-pass cooling water?
    - If yes, volume (gallons per day)?
    - Where is the discharge point?

11. Any other cooling water used? Describe...

12. Is there any water reuse in the plant? Any recycling of water?
    - If yes, where and what is it?

13. Is there any reverse osmosis or de-ionized water produced at this plant for incorporation into product?
    - If yes, is it on a service? Who maintains the equipment?
    - What, if any, pretreatment is there for the systems?...Describe?
    - What is the reverse osmosis reject ratio? (I.E., what percent of the reverse osmosis goes back to the sewer as reject and what percent is used as process water?)
    - Is water kept in a holding tank and drawn off as needed?
    - Or is it produced on an as-needed basis?
    - Is the water metered in? Out?

14. How much water (gallons per day) is incorporated into the product?

15. Do all process streams flow to a common interceptor point?
    - If yes, where is that point?
    - If no, where are the various points?
    - Is any single-pass or brine reject water co-mingling with regulated wastewater at the sample point(s)?

16. Is there any pretreatment of waste water? (Clarifiers, flocculation, etc.)
    - If yes, describe the various aspects in place and where located.
    - How is the bottom sludge handled?

17. How is the routine disposal of hazardous wastes handled?
    - Who is the waste hauler?
    - What is the average volume of material disposed to landfill, recycled, etc.?
    - Where are the records kept? (Can ask to see manifests.)

18. Is there any solvent reclamation done on-site?
    - Is any solvent collected for reclamation or recycling via return to the supplier?

19. Is there any paint recoup done at this site? Explain...

20. Is there any ink recovery and reuse at this site? Explain...

21. Is there adequate spill containment in the processing area(s)?
    - If yes, describe.
    - If no, describe.
    - How is the sewer otherwise protected from leaks or spills in the process area(s) and throughout the facility? Describe in detail.

Section G

PHARMACEUTICALS

INTRODUCTION

The pharmaceuticals category encompasses the manufacture of biological products, medicinal chemicals, botanical products and pharmaceutical products. The industry is characterized by diversity of product, process, plant size and process stream complexity.

This section will present summary information for the following topics:

- ❑ Definition of the Industry

- ❑ Manufacturing Processes

- ❑ Raw Materials and Products

- ❑ Effluent Guidelines

- ❑ Inspector Questions

❏ DEFINITION OF THE INDUSTRY

The pharmaceutical manufacturing category is grouped into five product or activity areas. This subcategorization is based on distinct differences in manufacturing processes, raw materials, products, and wastewater characteristics and treatability. The five subcategories are:

- Subcategory A - Fermentation Products

- Subcategory B - Biological and Natural Extraction Products

- Subcategory C - Chemical Synthesis Products

- Subcategory D - Formulation Products

- Subcategory E - Pharmaceutical Research - (not covered here)

<u>Subcategory D (formulating/mixing/compounding) is the most prevalent pharmaceutical manufacturing operation, with 80 percent of the plants in the industry engaged in this activity</u>. Fifty-eight percent of the plants have operations in only Subcategory D. The remainder also have operations in other subcategories.

❏ MANUFACTURING PROCESSES

One of the most important generalizations which can be made about the wastewaters produced and discharged by the pharmaceutical industry is their extreme diversity. Products, processes and the materials to which wastewater is exposed vary greatly. With the goal of relating those discharges with some common characteristics, subcategories based on unit manufacturing processes were defined. The broad manufacturing processing areas considered were:

- Fermentation

- Biological and natural extraction

- Chemical synthesis

- Formulation

Batch-type production is the most common type of manufacturing technique for each of the four subcategories.

One characteristic of processing in this industry is that the ratio of finished product to the quantity of raw materials is generally very low. This is most apparent in natural extraction (Subcategory B), followed by fermentation (A), synthesis (C), and formulation (D).

- **Fermentation**

    Wastewater from fermentation is generally high in BOD, COD, and TSS with a pH range of 4-8 units.

    Fermentation is the usual method for producing most antibiotics and steroids. The fermentation process involves three basic steps:

    - Inoculum and seed preparation
    - Fermentation
    - Product recovery

    Fermentation is conventionally a large-scale batch process. Sterilized nutrient raw materials in water are charged to the fermenter. Microorganisms are transferred to the fermenter from the seed tank and fermentation begins. After a fermentation period of from twelve hours to a week, the fermenter batch whole broth is ready for filtration. Filtration removes mycelia (remains of the microorganisms), leaving the filtered aqueous broth containing product and residual nutrients ready to enter the product recovery phase.

    There are three common methods of product recovery:

    - Solvent extraction
    - Direct precipitation
    - Ion exchange or adsorption

    Solvent extraction is a recovery process in which an organic solvent is used to remove the pharmaceutical product from the aqueous broth and form a more concentrated solution. The typical processing solvents used in fermentation operations are: benzene, chloroform, 1,1-dichloroethylene, and 1,2-trans-dichloroethylene.

    Direct precipitation consists of first precipitating the product from the aqueous broth, then filtering the broth, and finally extracting the product from the solid residues. Priority pollutants known to be used in the precipitation process are copper and zinc.

Ion exchange or adsorption involves the removal of the product from the broth, using such solid materials as ion exchange resin, adsorptive resin, or activated carbon. The product is recovered from the solid phase with the use of a solvent; it is then recovered from the solvent.

Steam is used as the major sterilizing medium for most equipment. However, to the extent that chemical disinfectants may be used, they can contribute to priority pollutant waste loads. Phenol is a commonly used disinfectant.

Sometimes a fermentation batch can become infested with a phage, a virus that attacks microorganisms. Usually the batch is discharged early and its nutrient pollutant concentration is higher than that of spent broth.

Another fermentation wastewater source is the control equipment that is sometimes installed to clean fermentation waste off-gas. The air and gas vented from the fermenters usually contain odoriferous substances and large quantities of carbon dioxide. Treatment is often necessary to deodorize the gas before its release to the atmosphere. Some plants employ incineration methods; others use liquid scrubbers. The blowdown from scrubbers may contain absorption chemicals, light soluble organic compounds, and heavier insoluble organic oils and waxes. However, wastewater from this source is unlikely to contain priority pollutants.

- **Biological and Natural Extraction**

　　　　Wastewater from biological and natural extractions is commonly lower in volume and mostly originates from the cleaning process areas since a very high degree of cleanliness must be maintained. Some solvent residues can be expected but the major waste is solid waste residuals of the original raw materials being used. Many materials used as pharmaceuticals are derived from natural sources such as plants, animal organs, and parasitic fungi. These materials have pharmaceutical applications ranging from tranquilizers to insulin. Also included in this group is blood fractionation, which involves the production of plasma and its chemical derivatives.

　　　　Despite this diversity, all extractive pharmaceuticals have a common characteristic: they are too complex to synthesize commercially. Extraction is an expensive manufacturing process since it requires the collection and processing of very large volumes of plant or animal matter to produce very small quantities of final product. In almost every step, the volume of material being handled is reduced significantly in each step of the process.

- **Chemical Synthesis**

    Wastewater from chemical synthesis is generally high in BOD, COD, and TSS with a pH range of 1-11 units. Most compounds used as drugs today are prepared by chemical synthesis. The basic major equipment item is the conventional batch reaction vessel. It is made of either stainless steel or glass-lined carbon-steel and contains a carbon-steel outer shell suitable for either cooling water or steam. The basic vessels may be fitted with many different attachments. Baffles usually contain temperature sensors to measure the temperature of the reactor contents. Dip tubes are available to introduce reagents into the vessels below the liquid surface. Typically, batch reactors are installed with only the top heads extending above the operating floor of the plant in order to provide the operator with easy access for loading and cleaning. Solutions can be mixed, boiled, and chilled in them. By addition of reflux condensation, complete reflux operations are possible. By application of a vacuum, the vessels become vacuum evaporators. Solvent extraction operations can be conducted in them, and, by operating the agitator at slow speed, they serve as crystallizers.

    Synthetic pharmaceutical manufacture consists of using one or more of these vessels to perform in a step-by-step fashion the various operations necessary to make the product. Following a definite recipe, the operator or a programmed computer adds reagents; increases or decreases the flow rate of cooling water, chilled water, or steam; and starts and stops pumps to transfer the reactor contents into another similar vessel. At appropriate steps in the process, solutions are pumped through filters or centrifuges or are pumped into solvent recovery headers or waste sewers.

    Each pharmaceutical is usually manufactured in a "campaign" in which one or more process unit is employed for a few weeks or months to manufacture enough compound to satisfy its projected sales demand. At the end of this campaign, the same equipment and operating personnel are used to make a completely different product, utilizing different raw materials, executing a different recipe, and creating different wastes.

    The synthetic pharmaceuticals industry uses a wide variety of priority pollutants as reaction and purification solvents. However, benzene and toluene are the most widely used organic solvents since they are stable compounds that do not easily take part in chemical reactions. Similar ring-type compounds (xylene, cyclohexane, pyridine, etc.) also are reported as being used in the manufacture of synthesized pharmaceuticals or resulting from unwanted side reactions.

Essentially all production plants operate solvent recovery facilities that purify contaminated solvent for reuse. Many of the wastes from the synthetic pharmaceutical industry will be discharged from these solvent recovery facilities. Aqueous wastes which may result form such operations include residues saturated with the solvents recovered.

The effluent from chemical synthesis operations is the most complex to treat because of the many types of operations and chemical reactions (nitration, amination, halogenation, sulfonation, alkylation, etc.). The production steps may generate acids, bases, cyanides, metals, and many other pollutants. In some instances, process solutions and vessel wash waters may also contain residual solvents.

Primary sources of wastewater from chemical synthesis operations are:

- Spent solvents, filtrates, concentrates, etc.

- Floor and equipment wash waters

- Wet scrubber spent waters

- Spills

Wastewaters from Subcategory C plant can be characterized as having high BOD, COD, and TSS concentrations; large flows; and extremely variable pH, ranging from 1.0 to 11.0.

- **Formulation**

    Although pharmaceutically active ingredients are produced in bulk form, they must be prepared in dosage form for use by the consumer. Pharmaceutical compounds can be formulated into tablets, capsules, liquids, or ointments.

    Tablets are formed in a tablet press machine by blending the active ingredient, filler, and binder. Some tablets are coated by tumbling with a coating material and drying. The filler (usually starch, sugar, etc.) is required to dilute the active medicinal to the proper concentration, a binder (such as corn syrup or starch) is necessary to bind the tablet particles together. A lubricant (such as magnesium stearate) may be added for proper tablet machine operation. The dust generated during the mixing and tableting operation is collected and is usually recycled directly to the same batch. Broken tablets are generally collected and recycled to the granulation operation in a subsequent lot. After the tablets have been coated and dried, they are bottled and packaged.

    Capsules are produced by first forming the hard gelatine shell. These shells are produced by machines that dip rows of rounded metal dowels into a molten gelatine solution and then strip the capsules from the dowels after the capsules have cooled and solidified. Imperfect empty capsules are remelted and reused, if possible, or sold for glue manufacture. Most pharmaceutical companies purchase empty capsules from a few specialist producers.

    The active ingredient and any filler are mixed before being poured by machine into the empty gelatine capsules. The filled capsules are bottled and packaged. As in the case of tablet production, some dust is generated. Although this is recycled, small amounts of waste dust must be disposed of. Some glass and packaging waste from broken bottles and cartons also result from this operation.

    Liquid preparations can be formulated for injection or oral use. In either case, the liquid is first weighed and then dissolved in water. Injectable solutions are bulk sterilized by heat or filtration and then poured into sterilized bottles. Oral liquid preparations may be bottled directly without the sterilization steps.

    Wastewaters are generated by general cleanup operations, spills, and breakage. Bad batches may create a solid waste disposal problem.

The primary objective of mixing/compounding/formulation operations is to convert the manufactured products into a final, usable form. The necessary production steps have typically small wastewater flows because very few of the unit operations use water in a way that would cause wastewater generation. The primary use of water in the actual formulating process is for cooling water in the chilling units and for equipment and floor wash.

Sources of wastewater from mixing/compounding/formulation operations are:

- Floor and equipment wash waters

- Wet scrubbers

- Spills

- Laboratory wastes

The use of water to clean out mixing tanks can flush materials of unusual quantity and concentration into the plant sewer system. The washouts from recipe kettles may be used to prepare the master batches of the pharmaceutical compounds and may contain inorganic salts, sugars, syrup, etc. Other sources of contaminated wastewater are dust and fumes from scrubbers either in building ventilation systems or on specific equipment. In general, these wastewaters are readily treatable by biological treatment systems.

Wastewaters from Subcategory D plants normally have low BOD, COD, and TSS concentration; relatively small flows; and pH values of 6.0 to 8.0.

❑ RAW MATERIALS AND PRODUCTS

The pharmaceutical industry utilizes a vast array of raw materials and processing agents. The diversity of feedstock is attributable to the variety of products and the number of process variations common to the industry.

Fermentation operations use large quantities of nutrient materials such as carbohydrates and proteins. Examples of some raw materials are meat extractions and distillers extract. Materials classified as priority pollutants which enter fermentation operations are mainly metals, as reaction modifiers and processing agents in the fermenter, and organic solvents, which are employed as extractive agents for product separation and purification. The residues form the organic starting materials plus mycelia contribute heavily to conventional BOD loadings.

Biological and natural extraction processes can have a wide variety of feedstocks including roots, leaves of plants, animal glands or parasite fungi. These substances contribute to BOD loadings; priority pollutant loadings are primarily due to solvents used for extraction. These solvents can be any number of organic compounds with benzene and chloroform being among the most widely used.

Chemical synthesis presents the broadest spectrum of starting materials. Feedstocks can range from oxbile to dextrose. Given the appropriate starting material there are many common synthetic processes (as many as several hundred) by which the starting material is transformed to the product. A number of solvents and additives are required to complete the synthesis. Solvents are usually inexpensive relative to the product and are used liberally for this reason. These solvents are almost exclusively organic and may be priority pollutants. Additives are used to control reactions and many contain metals that are priority pollutants.

In summary, chemical materials utilized and produced in the pharmaceutical industry are numerous and diverse. They are used as reactants, extractive solvents, catalysts, inhibitors, dilutents, and other purposes. In addition, other chemical compounds may be identified as intermediates, products, and byproducts. Many of these materials are among those listed as priority pollutants. In fact, the vast majority of the 126 priority pollutants listed are present somewhere in the industry although not necessarily in wastewater.

EFFLUENT GUIDELINES

FOR THOSE PLANTS USING OR GENERATING CYANIDE IN THE MANUFACTURING PROCESS, THE ALLOWABLE EFFLUENTS ARE SHOWN BELOW

Pollutant or Pollutant Property	BPT effluent limitations	
	Maximum for any one day	Average of daily values for 30 consecutive days
	Milligrams per liter (mg/l)	
Total Cyanide	33.5	9.4

If all cyanide-containing waste streams are diverted to a cyanide destruction unit and the effluent from the cyanide destruction unit is discharged to a biological treatment system, self-monitoring must be conducted after cyanide treatment and before dilution with other streams.

However, if all cyanide-containing waste streams are not treated in a cyanide destruction unit or if the effluent from the cyanide destruction unit is not discharged to a biological treatment system, self-monitoring must be conducted at the final effluent discharge point and the <u>daily maximum cyanide limitation must be multiplied by 0.35, and both limitations must be adjusted based on the dilution ratio of the cyanide-contaminated waste stream flow to the total process wastewater discharge flow</u>. Permittees not using or generating cyanide must certify to the permit-issuing authority that they are not using or generating this compound.

INSPECTOR'S QUESTIONS—PHARMACEUTICALS MANUFACTURING

1. What type of processes are used to manufacture product(s)?
   - fermentation
   - biological and natural extraction
   - chemical synthesis
   - mixing/compounding/formulation

2. If processes include fermentation and/or chemical synthesis, are these continuous or batch-type operations?

3. If chemical synthesis is involved, what processing steps produce wastewaters?
   - crystallization?
   - distillation?
   - filtration?
   - centrifugation?
   - vacuum filtration?
   - solvent extraction?

4. Are these wastewaters discharged to the sewer system?

5. What types of solvents are used, if any?

6. How are spent solvents disposed of?
   - Has Solvent Management Plan been submitted? Date submitted?

7. Is there any reverse osmosis or de-ionized water?
   - If yes, is it on a service? Who maintains the equipment?
   - What, if any, pretreatment is there for the systems?...Describe?
   - What is the reverse osmosis reject ratio? (I.E., what percent of the reverse osmosis goes back to the sewer as reject and what percent is used as process water?)
   - Is water kept in a holding tank and drawn off as needed?
   - Or do you produce on an as-needed basis?
   - Do you meter this water in? Out?
   - Where is the sample point for the wastewater discharge?

- Does regulated wastewater from the pharmaceutical operations also flow to this location?

8. Any requirements for steam generation or use? Describe...
    - If yes, is make-up water metered?
    - Volume (gallons per day)?
    - Is there blowdown to sewer? ...Frequency? ...Volume (gallons per day)?
    - What are the steam losses (gallons per day)?

9. Is any vapor degreasing done?
    - If yes, is the unit water cooled (one-pass), on a cooling tower or refrigerant type?
    - If one-pass, where does it discharge to sewer?
    - What type of solvent is used in the vapor degreaser?
    - How is spent or contaminated solvent handled? Describe...

10. What are the individual volumes and locations of the process tanks?

11. Any detergent cleaning or acid cleaning of tanks or vessels?
    - Other disinfecting chemicals used?
    - If yes, how frequently are these vessels batch discharged to the sewer?
    - What pretreatment occurs prior to discharge?
    - What are the individual volumes of all process vessels or tanks in facility?

12. How are spent or contaminated processing tanks or vessels handled?
    - If hauled, who hauls it? (Can check manifests.)

13. Are any tanks heated?
    - If yes, which ones?
    - How are they heated?

14. Any rectifiers, compressors or similar equipment in use?
    - If yes, do any of them require cooling water?
    - Any one-pass cooling water?
    - If yes, volume (gallons per day)?
    - Where is the discharge point?
    - If water is not used for cooling equipment, describe the cooling practices.

15. Is there any routine washdown of the work area(s)?
    - Where is the discharge point?

16. Are there any floor drains in the process area?
    - Does their flow pass through a common interceptor point?
    - If no, where does it discharge?

17. Do all process streams flow to a common interceptor point?
    - If yes, where is that point?
    - If no, where are the various points?
    - Is one-pass or brine reject water co-mingling with categorically regulated wastewater at the sample point(s)?

18. Is there any pretreatment of waste water?
    - If yes, describe the various aspects in place and where located.

19. How is the routine disposal of hazardous wastes handled?
    - Who is the waste hauler?
    - What is the average volume of material disposed to landfill, recycled, etc.?
    - Where are the records kept? (Can ask to see manifests.)

20. Is there any water reuse in the plant?
    - If yes, where and what is it?

21. Is there adequate spill containment in the processing area(s)?
    - If yes, describe...
    - If no, describe...
    - How is the sewer otherwise protected from leaks or spills in the process area(s) and throughout the facility? Describe in detail...

22. Has 'cyanide certification' been submitted? Date submitted?

Section H

PHOTO FINISHING

Introduction

All photo finishing activity consists of two main operations: developing film and/or printing paper. Film can be either **color** or **black and white** and the paper used corresponds to the type of film being processed.

Photographic processing consists of treating a silver halide-sensitized material with a series of chemical solutions and wash water steps to produce a visible image in black and white or color.

The final product is usually a color or black and white print, but 35mm slides or transparencies are also conventional products of photofinishing operations.

There are more than 20 different processes used with variations in the solution and sequence of the chemistry. Most facilities process a variety of materials and use more than one process. The newest convention is the Plumbingless Mini-Lab.

Processing may be accomplished either manually, meaning trays are used for the chemicals and rinsing operations, or via automated processors. The major sources of process wastewater are from (1) waste chemical solutions and (2) waste wash/rinse waters. The pollutants of concern are silver and, to a lesser degree, cyanide and chromium. The silver is contained in the emulsion of all the processed materials and is present in the waste water from all facilities. Cyanide and chromium are present in some bleach solutions in the form of ferricyanide and dichromate compounds, and are present only if these types of bleaches are used.

This section examines the following areas in greater depth:

- ❏ Black and White Processing

- ❏ Color Processing

- ❏ One-Step versus Two-Step Processing

- Water Uses
- Processing Equipment - Manual/Automated
- Mini Labs and Plumbingless Labs
- EPA Discharge Limits
- Pollutants of Concern
- Pretreatment/Silver Recovery
- Inspector's Questions

❏ Black and White Processing

The processing of black and white film and paper is a less critical chemical procedure than is the color procedure. The typical step are as follows:

- Developer

- Stop bath of either dilute acetic acid or water

- Fixer solution

- Optional hypoclearing agent

- Final rinse followed by placement on drying rack or drum

Black and white film is traditionally processed* with one of the six mainstay developers: D-76, HC-110, Microdol-X, Polydol, DK-50, or D-19. D-76, Microdol-X or Polydol are the most popular developers found in those shops which still do black and white processing. With the advent of automated 35mm cameras, automated photoprocessors and reasonably priced color film, the incidence of color photography - both professional and amateur - has risen dramatically. So much so that the black and white processing arena is primarily for special or artistic purposes, or as a hobby, and is not representative of the major commercial focus in photography today.

The steps in processing** the paper are very similar to the film processing procedure.

In contrast to the color process, black and white processing is not usually automated, although there is no reason why it could not be if the commercial and economic appeal existed. Instead, black and white is usually totally manual in its execution, with the major exception being large volume users of a black and white product such as newspapers. Even then it is usually only the film processing that is automated, not the printing portion.

* See Figure H-1
** See Figure H-2

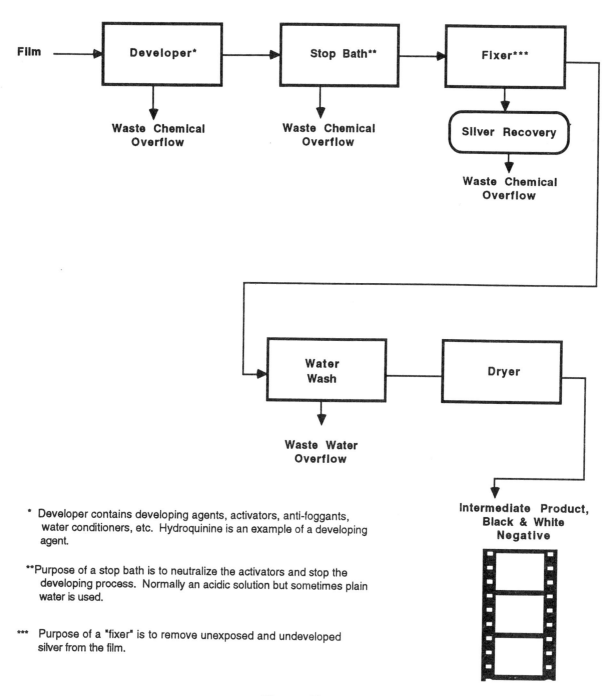

Figure H-1

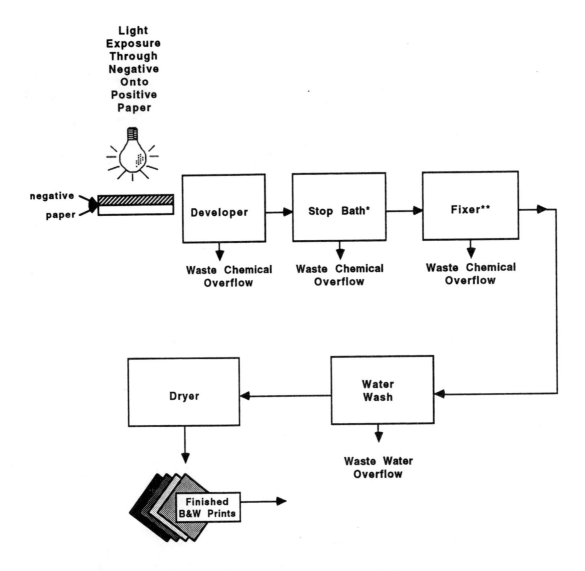

* STOP neutralizes the activators and stops the developing process

** FIXER or HYPOFIX removes the unexposed and undeveloped silver from the paper

Figure H-2

Black and White Reversal Film Processing

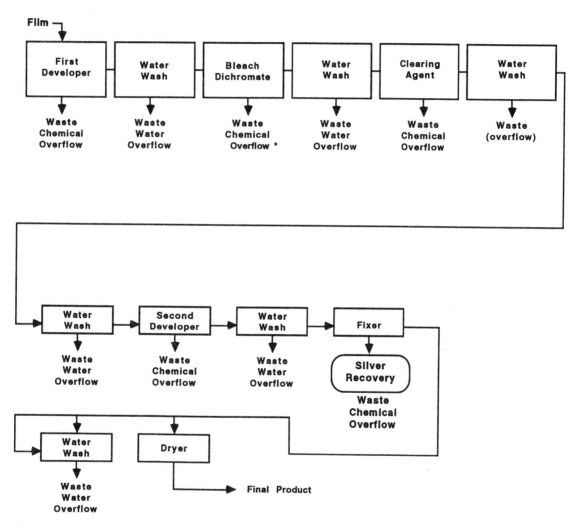

* Chromium is only present in wastewater from the black and white reversal film process.

Figure H-3

❑ Color Processing

The commonly used materials in color photographic processing are color negative film, color reversal film, color print film, and color print papers. The three basic processes for color materials are the negative process, the print process, reversal process with couplers in the emulsion (IC), or reversal process with coupler in the developers (DC). The commonly used color materials, uses, and process types are given in Table H-1.

Table H-1

Common Color Materials

Color Material	Use	Process	Image
Negative film	Original exposure, intermediate for copying positive transparencies	Negative	Negative
Positive print paper	Print from negative film	Negative	Positive
Reversal film	Original exposure, intermediate	Reversal (IC) or Reversal (DC)	Positive or Negative
Positive print film	Positive transparency from negative film	Negative	Negative
Reversal paper	Print from positive transparency	Reversal (IC)	Positive

In the five processes noted above, silver is always found in the wastewater. Chromium, however, is only present in wastewater from the <u>Black and White Reversal film process</u>. Cyanide is only found if certain ferricyanide bleaches are used in any of the color processes; rarely found in a Black and White procedure.

❏ Color Negative Process

For the production of a color negative transparency, the first step, color development, produces a dye image and a silver image in each layer of film. The amount of dye generated is proportional to the silver developed. (The images are negative with respect to the exposure sources.) The dye image is formed by a reaction between the developer oxidation products and a group of organic molecules called couplers which form dyes of the appropriate color in specific layers of the film.

The next step stops development and removes excess developer. This can be done by using either dilute acetic acid or a water wash. The film is then bleached to convert the developed silver image back to a silver halide in preparation for the subsequent removal of all silver from the final product by the fix solution. The color dye remains unaffected by these procedures.

After bleaching, the film is "fixed" to remove the remaining silver compounds, and water washed to remove the excess chemicals. Finally, the emulsion is hardened and stabilized using a weak solution of Formalin and sodium carbonate and passed through one final water wash before going to the drying phase.

See Figure H-4.

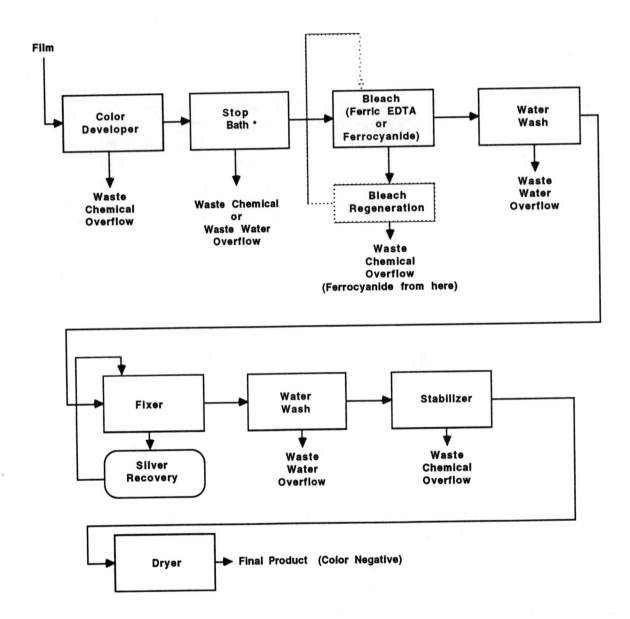

Figure H-4

❏ Color Printing Process

The color printing process is essentially the same as the color negative process in that the first step is color development. The color development produces a dye image and a corresponding silver image, the amount of dye generated being proportional to the silver developed.

The dye image is formed by a reaction between the oxidation products of the developer and a group of organic molecules called couplers to form dyes of the appropriate color.

The next step stops development and removes the excess developer using either dilute acetic acid or a water wash. The paper is then bleached to convert the developed silver image back to silver halide. After bleaching the paper is "fixed" to remove the remaining silver. Most color paper processes use a combination of a Bleach-Fix solution (BLIX), which converts silver to the halide form and dissolves the halides in one operation. Finally, the paper is washed to remove all traces of chemicals before proceeding to the drying phase.

Color Print Process

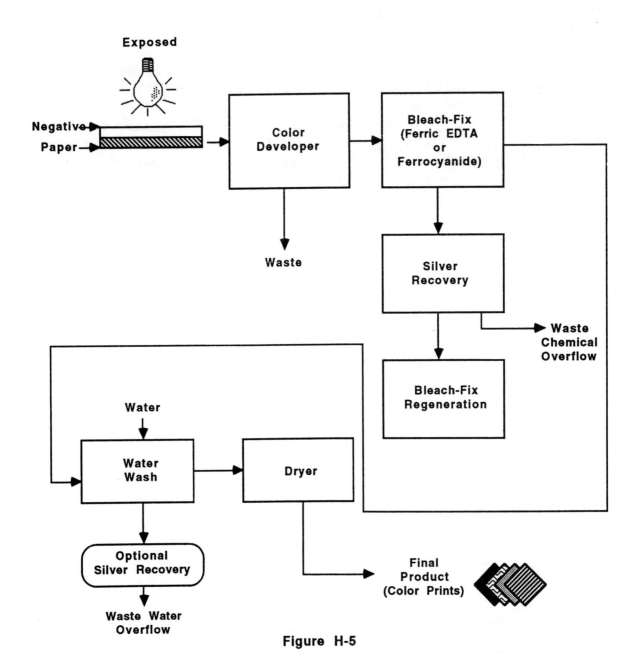

Figure H-5

❏ Color Reversal Processes

There are two different types of color reversal materials. In one, the color couplers which form the color dye image are incorporated into the emulsion layers at the time of manufacture (IC*). Most color reversal materials are of this type. The second type has three black and white layers, each sensitive to a different color. For this type of material, the color couplers are added during development (DC**). The (IC) process applies to film and paper materials, and the (DC) process applies to film only.

In the color reversal process (IC), a commonly used film bleach is potassium ferricyanide.

* See Figure H-6.

** See Figure H-7.

Color Reversal Film (E-6)*
Coupler in the Emulsion (IC)

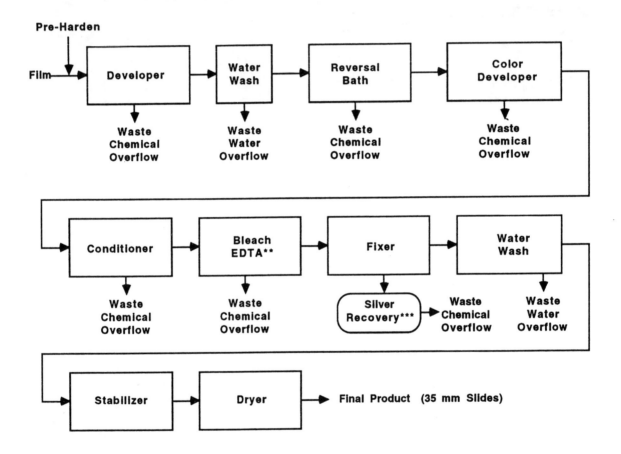

* EKTACHROME SLIDE PROCEDURE
** No regeneration of this bleach
*** No recycling of Fix because of bleach contamination

Figure H-6

Color Reversal Film (K-14)*
Coupler in the Developer (DC)

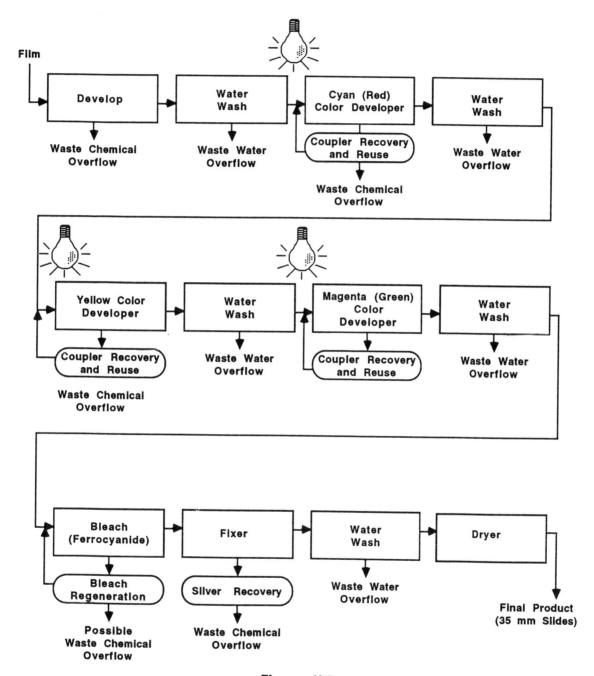

Figure H-7

* Kodachrome Slide Procedure

COLOR CHEMISTRY USED

C-41 ---------------------------------- Kodacolor 100, 200, 400 for Color Prints

K-14 ---------------------------------- Kodachrome 25, 40 (5070), 64 for Color Slides

E-6 ---------------------------------- Ektachrome 64, 160, 200, 400 Film for Color
                                                                         Slides; Ektachrome Slide Duplicating Film 5071

E-4 ---------------------------------- Ektachrome Infrared Film for Prints

Figure H-8

❑ One-Step Versus Two-Step Process

Either a one-step or a two-step procedure is used to produce a positive image of the subject on film or paper. In the <u>two-step procedure</u>*, an intermediate chemical process called the <u>"negative" process</u> is used. (The negative process is utilized for producing either a negative or a positive image on film or paper.) The first step of the negative process consists of producing a negative image on film, called a negative transparency. The transparency is then used as a light mask in the second step to expose paper or film. In the second step, the material is processed by the specific chemical process dictated for the type of film and/or paper being used.

In contrast, the one-step procedure** uses a <u>"reversal"</u> chemical process which directly produces a positive image of the subject on the film or paper used. This process is most often used for the production of motion picture films and 35 mm (projection) slides. Also, certain black and white materials, commonly referred to as reversal products, yield a direct "reversal" (positive) image. A dichromate bleach (<u>which is not regenerated</u>) is frequently used in the black and white reversal processing. It is this black and white reversal procedure which is the usual source of chromium found in these particular wastewater discharges.

* See Figure H-3.

** See Figure H-4 or Figure H-5.

❏ Water Uses

Water is used in the photographic processing industry for the following purposes:

- As process water for film and paper wash water, solution make-up water, and area and equipment wash water

- As non-process water as in non-contact cooling water (subject to local ordinances), and heating water and/or cooling towers.

The largest, single process water use is for washing of film and paper during various stages of the product development. The function of the wash step is to remove chemicals included in the emulsion during manufacture (but no longer needed), absorbed into the emulsion during processing, or reaction products created during processing. 95% of the process water used in each facility is for film and paper washing.

The chemicals used to make up processing solutions are generally supplied to the processor in the form of liquid concentrates or powdered chemical formulations. The processor adds water to make up the solutions to desired strength. Wastes are generated when these solutions are discarded after becoming exhausted or when allowed to overflow during replenishment. Both situations are common practice.

In addition to the above stated uses, water is commonly used for the washing and rinsing of solution mixing utensils, storage tanks, processing machines and for area washdown.

Some facilities use non-contact water for cooling of process solutions to maintain proper operating temperatures. This water is not process water since it is contained in enclosed water jackets around process tanks and does not come into contact with raw materials or the product.

Non-contact cooling water presents two problems. This water constitutes dilution and must be factored out when limits are calculated. For example, if the cooling water is 10,000 gallons per day and the process water is also 10,000 gallons per day the flow is 50% diluted by clean, non-contact cooling water. It follows that all of the permit limits are also reduced by 50% (e.g., 2.0 mg/l becomes 1.0 mg/l).

Additionally, non-contact cooling water usually violates the local sewer ordinance which strictly prohibits the discharge of clean water. Heavy cooling water demands at photo shops are more appropriately supplied by recirculating water from cooling towers (see Cooling Tower Section).

❏ Processing Equipment - Manual/Automated

Photographic materials may be processed manually or in automatic processing machines. Manual processing, also referred to as "tray processing" or "sink line processing", consists of manually placing photographic materials in each of a series or trays (for exposed paper), and tanks (for exposed film) which contain specific chemical solutions or water rinses. This processing technique is usually used for low volume operations or specialty work.

High volume processing, however, is done in automated processors. In an automatic film processor, multiple rolls of film are simultaneously and continuously loaded into a machine and the film travels through a series of chemical containers vial spools and rollers. This film undergoes reactions, emerging at the end of the process fully developed and dried.

From the automatic film processor, the material can be set up into a slide format, if the process was color reversal. Or, the film could be loaded onto an automated print processor. Multiple rolls of film are sequentially loaded into the autoprocessor which feeds a continuous roll of unexposed print paper into contact with the film. The film and paper come into contact inside the processor, are exposed to light, and the exposed paper then proceeds through a series of chemical containers and undergoes development. The print emerges at the end fully processed and dried, cut, and ready for placement into an envelope along with the related negatives.

❑ Mini-Labs/Plumbingless Mini-Labs

A **conventional minilab** uses wash water in the same manner as other processors, and may discharge as much as 20 to 25 gallons of effluent for every roll of film processed. A "plumbingless" minilab, which uses a stabilizer in place of wash water, will discharge less than 0.1 gallon of effluent for each roll of film processed. For example, a conventional minilab that processes 50 rolls of film each day could generate 1,000 gallons of effluent. Most of the effluent is water. A "plumbingless" minilab processing the same amount of film will discharge less than 5 gallons of effluent each day.

Plumbingless minilabs that use Process RA-4NP and C-41BNP processing solutions offer several environmental advantages. The overall chemical loading with these new processes is actually less than that of minilabs using Processes EP-2, EP-2NP, C-41, and C-41NP. Also, the total volume of waste discharged for treatment is reduced by as much as 98 percent. Plumbingless minilabs conserve water, and costs for energy to heat the water are eliminated.

Typical Effluent Characteristics from Minilabs (Plumbingless) Using processes RA-4NP and C-41BNP

	40 rolls/day	100 rolls/day
Silver		
Recovery with pH adjustment	0.05 g/day (0.0001 lb/day)	0.11 g/day (0.00025 lb/day)
Recovery without pH adjustment	0.22 g/day (0.0005 lb/day)	0.55 g/day (0.0013 lb/day)
No recovery	22 g/day (0.049 lb/day)	56 g/day (0.12 lb/day)

Capitol Investment *

Recovery Method	100 rolls/day	1,000 rolls/day
Chemical Recovery Cartridges	$1000	$3000
Electrolytic (with tailing CRCs)	$2,500	$22,000
In-Situ Ion Exchange	$5000	$33,000

Typical Silver Concentrations in Effluent After Recovery

Recovery Method	Mg/L
Chemical Recovery Cartridges (CRCs)	10 to 20
Electrolytic (with tailing CRCs)	1 to 5

* A small shop might want to practice "saving fix" for pickup as the most economically desirable method of silver recovery/pretreatment

Figure H-9

❏ **EPA Discharge Limitations**

<u>Photographic processing industries producing less than 150 sq. m/day (= 1,600 sq. ft./day) are exempt from EPA discharge limitations.</u>

The industry average is 4,000 gallons/1,000 sq. ft. of production but flows of 50,000 GPD are not uncommon.

❑ **Pollutants of Concern**

The two major sources of waste during photo processing are:

- Overflow during replenishment of chemical solutions

- Rinse water overflow (which may be continuous)

The pollutants of significance are silver, cyanide and chromium. Most photo processors practice silver recovery and/or bleach or bleach-fix regeneration. This practice mitigates the toxicity of their waste water. It is not unusual, though, to find a very low-volume discharger who does not. Specifically in the instance of silver recovery, it is sometimes too difficult to retain the services of a silver recoverer, or too costly to install and maintain a traditional silver recovery system.

❑ Pretreatment/Silver Recovery

Two photoprocessing waste solutions contain essentially all of the silver that is leached out during the film and/or print processing:

- The fix or the bleach-fix (BLIX) overflow and/or spent solutions

- The post-fix rinse/wash water

The silver from the bleach-fix (BLIX) or fix solutions is generally recovered by (1) collecting and recycling the spent liquid chemicals to a reprocessing/silver recovery firm, (2) using a metallic replacement recovery technique, (3) installing an electrolytic silver recovery system, or (4) ion exchange. Some combination of the above methods may also be used.

Even with an efficient fix solution silver recovery system and squeegee use at the exit of the fix tank, up to 10 percent of the available silver can be lost to the after-fix wash water by carryover. The silver concentration in the wash water is typically in the range of 1 to 50 mg/l and it has been found that recovery of this silver is not as responsive to electrolytic or metallic replacement recovery methods. Nonetheless, many large photofinishing firms still use, and appear to benefit from, electrolytic recovery systems.

1) Collecting Spent Chemical Fixers

Spent chemical fixer are generally saved in containers which the recycler/reprocessor supplies. These can be plastic drums, a cube, or similar container. The recycler picks up the spent fixers from the facility and transfers the material to his own recycling site for the recovery process.

2) Metallic Replacement Method

Metallic replacement occurs when a metal, such as iron, comes into contact with a solution containing dissolved ions of a less active metal, such as silver. Because of its economy and convenience, iron in the form of steel wool is used most often. Silver recovery by metallic replacement is most often carried out using commercially available units consisting of a steel wool filled plastic bucket with appropriate plumbing. Typical practice is to feed waste fix to two or more canisters in series, the first canister removes the bulk of the silver and the second unit polishes the effluent. Silver concentrations in the effluent from a single unit average 40 to 100 mg/l over the life of the unit versus a range of 0.1 to 50 mg/l when two canisters are used in series.

3) Electrolytic Silver Recovery Method

Electrolytic silver recovery is the only method that permits fixer reuse. In this method, the silver-bearing solution is passed between two electrodes through which a controlled direct electric current flows. Silver plates out on the cathode as almost pure metal. Considerable agitation and large plating surface areas are necessary to achieve good plating efficiency and high quality silver up to 96-98 percent pure. The cathodes are removed periodically, and the silver is stripped off.

The advantage of the electrolytic method is that silver is recovered in an almost pure form, making it easier to handle and less costly to refine. With careful monitoring, it also permits fixer reuse for some processes. It also avoids the need to store and replace cartridges, as with the metallic replacement method.

Electrolytic units can be used for primary or tailing silver recovery. Primary electrolytic systems are typically installed in two basic ways. One is a batch recovery system where overflow fix from a process line or lines is collected in a tank. When sufficient volume is reached, the waste fix is pumped to an electrolytic cell for the silver removal process. The silver concentration in the effluent is typically about 200-500 mg/l but can be reduced to 20-50 mg/l with additional treatment time and careful control of current density.

The second primary electrolytic recovery method is to remove silver from the fix solution from one or more process machines in a continuously recirculating system at approximately the rate at which silver is being added by processing. The recovery cell is included "in-line" as part of the recirculation system.

4) Ion Exchange

(See Treatment Technologies for complete details on this procedure.)

INSPECTOR'S QUESTIONS—PHOTOFINISHING

1. What types of photo processing are done at this facility?
    - Black and white film?
    - Black and white prints?
    - Black and white reversal?
    - Color film?
    - Color prints?
    - Color reversal?

2. Are automated processors used or is work done manually? Or a combination of the two methods?
    - If work is done manually, identify which processes they are and describe describe the manual method, i.e., "dip and dunk", trays, etc., and the approximate volume of work produced or the length of time operation is in use per day in order to establish the waste water flow.

    - If work is processed automatically, identify which processes are automated (film, paper), how many autoprocessors there are and describe them (E-6 processors, etc.). Identify the gallons per minute throughput of each processor and the hours used per day. If this information is unavailable, estimate the approximate volume of water used per day, either via manufacturer's specifications, engineering estimates of pipe diameters or according to the water bill.

    - If work is processed both manually and with autoprocessors, use a combination of actual flow calculations using gallons per minute throughput, flow estimation, etc.

3. Does the rinse water from the processors or in the manual rinse operation run continuously, or is it shut off when no work is being processed? Describe.

4. How often are the processors or manual tanks/trays cleaned? The chemistry

changed? Describe.

5. What chemicals, if any, are used to clean the autoprocessor roller, tanks, trays, etc.? Describe.

6. What brand of chemistry is used (Kodak, 3M, GAF, etc.)? Approximately what volume is used per month? Per year?

7. What different types of process chemistry are used (E-6, C-41, etc.)? Describe.

8. What are the names of the chemicals used in the process? Describe.

9. Does any of the chemistry used contain cyanide?
 - If yes, is it discharged to the sewer? Where? What is the volume?

10. What volume of chemistry is discharged to the sewer? (Approximate volume per day/month, etc.)

11. Is bleach regeneration practiced and if so, is it done within the lab?
 - If yes, describe the processes and the wastes involved.

12. Is silver recovery practiced?
 - If yes, describe the recovery type (electrolytic, save fix, etc.).

13. Does any of the process equipment require cooling water?
 - Any one-pass cooling water?
 - If yes, volume (gallons per day)?
 - Where is the discharge point?

14. Is there any routine washdown of the work area(s)?
 - Where is the discharge point?

15. Are there any floor drains in the process area? In the chemical mixing and/or storage area?
 - Does their flow pass through a common interceptor point?
 - If no, where does it discharge?

H - 18

16. Do all process streams flow to a common interceptor point?
    - If yes, where is that point?
    - If no, where are the various points?
    - Is there any one-pass cooling water co-mingling with regulated waste-water at the sample point(s)?

17. How is the routine disposal of hazardous wastes handled?
    - Who is the waste hauler?
    - What is the average volume of material disposed to landfill, recycled, etc.?
    - Where are the records kept? (Can ask to see manifests.)

18. Is there any de-ionized water production?
    - If yes, is it on a service? Who maintains the equipment?
    - What, if any, pretreatment is there for the systems?...Describe?
    - Is water kept in a holding tank and drawn off as needed?
    - Or is it produced on an as-needed basis?
    - Is the water metered in? Out?

19. Is there any water reuse in the facility?
    - If yes, where and what is it?

20. Any requirements for steam generation or use?
    - If yes, is make-up water metered?
    - Volume (gallons per day)?
    - Is there blowdown to sewer? ...Frequency? ...Volume (gallons per day)?
    - What are the steam losses (gallons per day)?

21. Is there adequate spill containment in the processing area(s)? Describe...

22. Other than with spill containment, how is the sewer protected from leaks or accidental spills in the process area(s)? Explain...

Section I

PRINTING/PUBLISHING

Introduction

The printing/publishing industry group, as a whole, is involved in converting printed material and/or images from some prior format onto paper or other media. The gamut of printing procedures ranges from printing activities that use no chemicals and produce no waste to other activities which use many chemicals and produce waste of many different kinds. Some of these waste products are hazardous and contribute to an industrially oriented effluent which requires permitting and regular monitoring.

The typical print shop which would be classified as "dry" would utilize computers, xerox machines or similar equipment to do all of their printing. Also, older facilities, or specialty printing shops, still use letterpress printing, a process which molds the individual characters out of lead. These specialty shops produce checks, letterheads, call-cards (business cards), and various business-related forms. The type is formed from molten lead in typographic machines (which may have water-cooled molds). The type is set up, put in brackets and placed into a printing press. Ink is rolled onto the type, type is contacted with the paper and the printed article is thus produced. Solvents can be used to clean the type, presses, etc., but this is usually rag applied only. When printing is completed, the type can be saved or smelted to produce fresh ingots for reuse. Generally speaking, there is no wastewater discharge from either of these type facilities. The focus of this section will be on silkscreening and off-set printing processes which could produce a permittable effluent. Discussions will include:

- ❏ Description of Silkscreening

- ❏ Description of Off-Set Printing

- ❏ Description of Captive Printing

- ❏ Water Uses

- ❏ Constituents of Concern

- Pretreatment
- Inspector's Questions

❑ Description of Silkscreen Printing

In this procedure, paint or ink is forced through a stencilled screen which allows only the positive image to be printed. Silkscreening is used primarily for reproducing large images onto a variety of materials ranging from wood and paper to clothing items such as T-shirts and baseball caps.

The stencilled screen is prepared by applying photosensitive liquid or paper onto a screen. The desired image is then exposed, with light, through a negative onto the screen. A developer is used to remove the non-exposed areas.

After the printing process is completed, the excess ink is usually cleaned off of the screen with some type of solvent thinner. If there is no reason to save the stencilled screen, the old stencil can be removed with detergent, and the screen reused for a new stencil.

❏ Description of Off-Set Printing

The procedure called "off-set" printing can be divided into three main segments as follows:

- Exposure of the image onto a black and white negative followed by development of the negative

- Exposure and development of the image (from the negative) onto a photosensitive plate

- Transfer of the image from the plate onto paper

All three stages are usually performed by a single firm, but could be "farmed out" to different specialty groups. (See Figure I-1.)

• Negative Developing/Processing

Exposing the images in the off-set procedure is normally done with a large stationary camera using 11" X 14" negative film. The remaining film development process is somewhat similar to the traditional black and white technique. The negative emulsion contains photoreactive silver halide crystals which are exposed to light, developed to convert crystals to silver, etc. Generally, there is no cyanide discharge from this procedure but occasionally some firms will use potassium ferricyanide (Farmer's Reducer) to further reduce some negatives. However, this should never amount to more than an ounce or so per day.

• Plate Exposure/Processing

The finished negative is then used to expose the original image onto a photosensitive plate. The plates are usually made up of aluminum with an overlaid organic photosensitive layer. The photosensitive layer is activated by a bright arc lamp and then a subtractive plate developer is used to remove the non-activated areas. What remains is the positive image which is also ink adsorbing.

Certain printed materials, which do not require long runs, may utilize paper plates rather than the aluminum type. Processing is very similar to the traditional black and white photo procedure but <u>cyanide may be present</u> in the chemical activator used.

- **Transfer Procedure/Presses**

Once completed, the plate is put onto a press roller and printing can commence. The plate is contacted with fountain solution through a series of rollers which apply a thin layer of solution onto the finished plate. The fountain solution typically contains water, alcohol, and special fountain concentrates. (Non-alcohol systems use a fountain concentrate which may contain either n-propyl alcohol or chromates.)

The fountain solution wets the plate except for the hardened photo activated areas which repels water. The plate is then contacted with a thin layer of ink through a series of rollers. The activated areas attract the grease in the ink while the water on the rest of the plate repels the ink. The plate is then contacted with the blanket roller, a vinyl-like sheet which picks up the ink and transfers it to the paper. In this fashion the printing continues and colors can even be changed by changing to a different press (each press contains only a single color: red, blue, yellow or black) - thereby producing multicolored printed material.

❑ Captive Printing Operations

In some industries where printing is not the main function, captive printing operations may be found. One of the more common is the box manufacturing industry. Many box companies will print the customer's name, product and other information on boxes. Offset printing is one of the methods commonly used for non-cardboard containers such as ordinary food boxes. The process and chemistry employed are similar to that of offset printing on paper, described earlier in this chapter.

Printing on cardboard can be done using rubber plates or a stencil. With rubber plate printing, ink is applied on to the surface of the plate by rollers and then contacted with the box. The plates are normally made by outside firms and can be used many times. Stencil printing consists of brushing ink manually or automatically through a paper or metal stencil onto the box. Wastes generated from box printing would come primarily from cleanup and spills. Where water soluble inks are used, water and detergent are applied to clean rollers, ink trays, plates, etc. with waste water being discharged to the sewer. For solvent-based inks, a petro-solvent similar to blanket wash would be used and should not be discharged to the sewer.

Generally any large company may have their own printing operation to fulfill production requirements. Large insurance companies, banks or science and engineering firms may print their own forms, checks or technical matter. Manufacturing firms may print their own labels, packaging, advertising literature, etc. Variations of the printing methods discussed are possible but the basic principles will be the same.

❑ Water Uses

As with similar photographic operations, a large part of the water used in printing is for washing of film and plates during plate production. In addition, within the printing industry, some percentage of their water use can go to cleaning of the actual equipment used. The exception is if all cleaning is solvent-based and rag applied only. Whether a printer is using water for cleanup or other solvents will depend largely on what type of printing is being done and the inks being used. E.g., water-based inks could be cleaned up with water, whereas oil-based inks would require the use of solvents.

Film processing may be done either manually or via automatic processors with high volume printers opting for the automated alternative.* During normal processing operations the rinse water from automatic processors is running and continues to run for about 5 minutes after processing is completed. In some older automated processors, however, there is no automatic shut off and water continues to run even after processing is completed. (Fortunately, these machines are being phased out and replaced by newer, more environmentally sound equipment.).

Plate processing utilizes a developer applied to the photo sensitive layer of the aluminum plate. Developers can be applied and rinsed off in a sink and/or wiped off with a rag. Also, some printers could have automatic plate developers which may or may not have an associated rinse step. If there is a rinse step, it is not uncommon to find that the rinse runs continuously, whether plates are being processed or not. (Essentially producing large volumes of uncontaminated water during low output times.)

Equipment cleanup can be accomplished with simple water cleanup, using detergents or other additives, strong caustic solutions and the like, with an associated wastewater discharge. On the other hand, many printers use no water in their cleanup procedures, utilizing only some type of solvent like Stoddard Solvent with added aromatic petroleum distillates (called Blanket Wash) to clean the blanket rollers, other rollers and the ink trays. Normal practice is to wipe up the solvent residuals with rags. In some shops, the solvent is poured onto the various rollers, ink trays, etc. during cleaning and collected back

* See Photo Section for complete details on film processing

after use. The only suitable repository for this spent solvent is a hazardous waste drum. Nonetheless, it is not unusual to have this material find its way into the rag barrel (and from there on to the industrial launderer), the trash dumpster, or even the sewer.

Occasionally, it may be that a very high speed press requires cooling water. This may be single-pass or a cooling tower may be in service for this purpose.

❑ **Constituents of Concern**

The two primary sources of pollution would be from (1) the production of the actual plate and (2) the cleanup of printing equipment, presses, rollers, ink trays, fountains, etc.

During plate production, the constituents of concern are silver (from the negative development process) and, to a much lesser degree, cyanide, from the same process.

If water soluble inks are used, there could be trace levels of heavy metals in the cleanup wastewater. It is more typical that oil-based inks are used and no "normal" cleanup fluids or solutions are discharged to sewer.

❑ **Pretreatment/Silver Recovery**

The wastewater process stream that contains all of the recoverable silver is from the film development sequence (see Figure I-1) and within that process it is specifically the fixer which has the significant concentration of silver. The printing plant operator has all of the same options open to him as the traditional photo processor, where silver recovery is concerned. However, the reality is that whether silver recovery is practiced or not depends largely on the volume of waste fixer generated and the concentration of silver in it. The simplest method for low volume recovery is to collect the fix and have it reclaimed by an outside firm rather than invest in expensive pretreatment equipment which may not have the necessary payback potential. Irrespective of the treatment method utilized, it is important that the inspector evaluate the site and include silver pretreatment/recovery to assure that excessive levels of silver do not pass into the sewerage system.

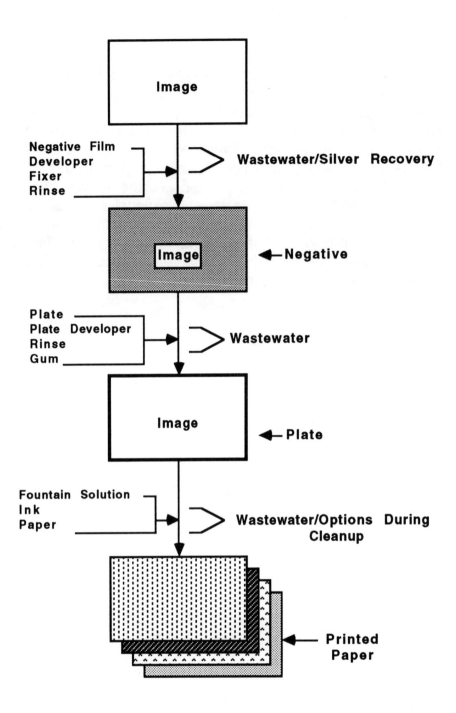

PROCESS FLOW - OFF-SET PRINTING

Figure I-1

I-10-a

INSPECTOR'S QUESTIONS—SILKSCREENING

1. What type of photo sensitive coating is used?
    - Paper?
    - Liquid?
    - If liquid, what volume?
    - Is it discharged to sewer? Explain...

2. What type of developer is used?
    - What volume?
    - It is discharged to the sewer?
    - If yes, where?

3. How is screen cleaned after printing?
    - If a solvent is used, what is its destination?
    - Sewer?
    - Hauled away? Explain...

4. Are the screens reused or are they thrown away? Describe...

5. Any rectifiers, compressors or similar equipment in use?
    - If yes, do any of them require cooling water?
    - Any one-pass cooling water?
    - If yes, volume (gallons per day)?
    - Where is the discharge point?

6. Is there any routine washdown of the work area(s)?
    - Where is the discharge point?

7. Are there any floor drains in the process area?
    - Does their flow pass through a common interceptor point?
    - If no, where does it discharge?

8. Do all process streams flow to a common interceptor point?

- If yes, where is that point?
- If no, where are the various points?

9. Is there any pretreatment of waste water?
    - If yes, describe the various aspects in place and where located.

10. How is the routine disposal of hazardous wastes handled?
    - Who is the waste hauler?
    - What is the average volume of material disposed to landfill, recycled, etc.?
    - Where are the records kept? (Can ask to see manifests.)

11. Is there any water reuse in the plant?
    - If yes, where and what is it?

INSPECTOR'S QUESTIONS—OFF-SET PRINTING

1. What types of printing are done in addition to off-set?
   - Letterpress?
   - Silkscreen?
   - Other types?

2. When the off-set printing is done, does this include:
   - Film processing?
   - Plate developing?

3. Are automated processors used or is work done manually? Or a combination of the two methods?
   - If work is done manually, identify which processes they are and describe the manual method, i.e., "dip and dunk", trays, etc., and the approximate volume of work produced or the length of time operation is in use per day in order to establish the waste water flow
   - If work is processed automatically, identify which processes are automated (film, paper), how many autoprocessors there are and describe them (E-6 processors, etc.). Identify the gallons per minute throughput of each processor and the hours used per day. If this information is unavailable, estimate the approximate volume of water used per day, either via manufacturer's specifications, engineering estimates of pipe diameters or according to the water bill.
   - If work is processed both manually and with autoprocessors, use a combination of actual flow calculations using gallons per minute throughput, flow estimation, etc.

4. Does the rinse water from the processors or in the manual rinse operation run continuously, or is it shut off when no work is being processed? Describe.

5. How often are the processors or manual tanks/trays cleaned? The chemistry changed? Describe.

6. What chemicals, if any, are used to clean the autoprocessor roller, tanks, trays, etc.? Describe.

7. What brand of chemistry is used (Kodak, 3M, GAF, etc.)? Approximately what volume is used per month? Per year?

8. What different types of process chemistry are used? Describe...

9. What are the names of the chemicals used in the process? Describe.

10. What volume of chemistry is discharged to the sewer? (Approximate volume per day/month, etc.)

11. Does any of the chemistry used contain cyanide?
- If yes, is it discharged to the sewer? Where? What is the volume?

12. If plate developing is done, what type of plates are used? Explain...

13. If they are aluminum plates, are they developed with...
- Subtractive developer?
- Color key developer (transfer developer)?
- Additive developer ("R" developer)?
- Explain...

14. Is the developer washed off to the sewer or wiped off with a rag? Explain....
- About how many plates are developed?

15. If paper plates are used, what type processor is used?
- What type of chemicals?
- What volume is used?
- What volume to sewer?

16. What type of fountain solution is used?
- What volume is used?
- What volume is discharged?

17. What type of solvent is used to clean:
    - the presses?
    - the rollers?
    - the fountains?
    - Explain...

18. Is this solvent discharged to sewer? Explain...

19. Is there any routine washdown of the work area(s)?
    - Where is the discharge point?

20. Are there any floor drains in the process area? In the chemical mixing and/or storage area?
    - Does their flow pass through a common interceptor point?
    - If no, where does it discharge?

21. Do all process streams flow to a common interceptor point?
    - If yes, where is that point?
    - If no, where are the various points?

22. Is silver recovery practiced?
    - If yes, describe the recovery type (electrolytic, save fix, etc.).

23. Does any of the process equipment require cooling water?
    - Any one-pass cooling water?
    - If yes, volume (gallons per day)?
    - Where is the discharge point?

24. How is the routine disposal of hazardous wastes handled?
    - Who is the waste hauler?
    - What is the average volume of material disposed to landfill, recycled, etc.?
    - Where are the records kept? (Can ask to see manifests.)

25. Is there any water reuse in the facility?
    - If yes, where and what is it?

26. Is there adequate spill containment in the process area(s)?
    - If yes, describe.
    - If no, describe.
    - How is the sewer otherwise protected from leaks or spills in the process area(s) and throughout the facility? Describe in detail.

Section J

STEAM PLANTS/BOILERS

Introduction

The principle reason for the existence of a steam power plant or a boiler is for heat generation from coal, oil or some other source of fuel. The end product of the heat is always steam.

Often, this steam is used to power turbines for the production of electricity. On a smaller scale, some industrial plants produce enormous quantities of steam simply for process uses (e.g., steam tunnels at commercial/industrial laundries).

This section will further develop the following areas as they apply to steam plants and/or boiler systems.

- **Steam Plants**
- **Industrial Boilers/Utility Boilers**
- **Water Treatment/Blowdown**
- **Water Use**
- **Return Condensate**
- **Pollutants of Concern**
- **Inspector's Questions**

❑ Steam Plants

Considering the many uses for energy today in industry, and for domestic applications, by far the greatest energy use is directed to steam production. Steam production occurs via the combustion of fossil fuels and, in some instances, from nuclear fission. The predominant users are utility companies, but large industrial plants account for a sizeable portion of steam production as well.

Usually, the steam produced is used to drive large turbines, in association with generators, whereby electrical energy is produced. Whether steam plants are located at utility companies or at large industries, they share one trait in common - <u>a large volume of water use</u>. This is mainly for cooling purposes at power generating stations. The second notable use of water is the actual water needed to make the steam.

This second type of water use applies equally to large or small steam plants, in that the amount of steam produced is directly proportional to the amount of water used, more or less.

❏ Industrial Boilers/Utility Boilers

The standard types of boilers found in industrial settings are of two types: (1) firetube and (2) watertube.

- **Firetube Boilers** confine the flame and hot gases within multiple tubes arranged in bundles (see figure J-1) which are enclosed by a water jacket. The water circulates around the exterior of these tubes. As the water heats up to steam temperature, it escapes at the top through the steam outlet pipe. Firetube boilers are used for application where the steam requirements are below 150,000 lbs per hour and 150 psi (this is referred to as a low pressure boiler system).

Firetube boilers are rated in terms of BOILER HORSEPOWER (one boiler horsepower equals 34.5 lbs of water evaporated at $212°$ F). These boilers are usually available in a horsepower range of 5 to 750 HP (producing about 25,000 lbs/hrs of steam), and at pressures up to 250 lbs/sq. in. This type of boiler does not come equipped with a superheater and is thus limited to the production of saturated steam only. They are usually fired by oil and gas-fuel only.

- **Watertube boilers** are usually of the oil- and gas-fuel firing type but can be coal-fired as well. Higher pressure and a greater capacity require the use of a watertube boiler. Watertube boilers are equipped with thicker, stronger plates and tube walls to withstand the higher temperatures and pressures produced. Combustion of the fuel occurs in a furnace and additional water tubes are commonly found in the furnace walls. This condition greatly enhances the boiler efficiency. The main difference between the watertube and the firetube is as follows: In the watertube boiler (see Figure J-2) the flame and hot gases flow across the outside of the tubes and water circulates within the tubes, whereas in the firetube the flame and hot gases are in the tubes and the water circulates outside the heated tubes.

Watertube boilers are usually furnished as shop built (a package) in the 10,000 to 250,000 lb/hr. size range and mostly are the oil- and gas-fuel firing type. The larger capacity oil- and gas-fuel and generally all of the coal firing units are field-erected. Many older watertube boilers are of the straight watertube construction but modern units are almost always of the bent watertube

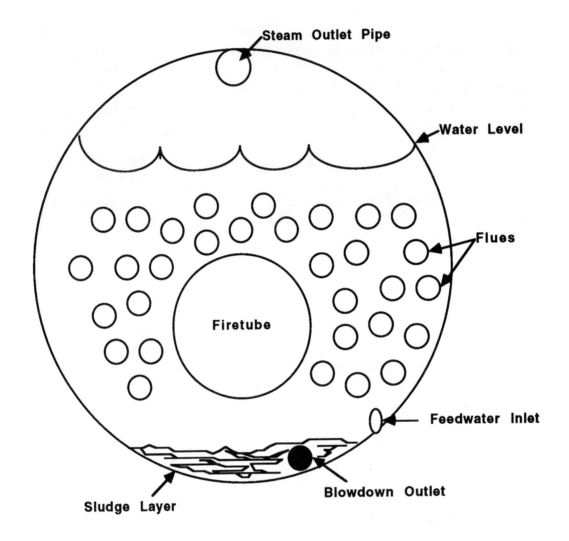

FIRETUBE BOILER

Figure J-1

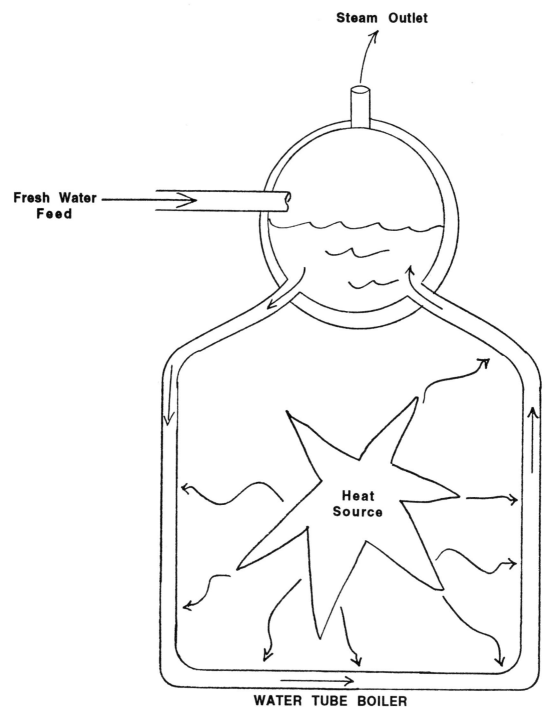

WATER TUBE BOILER
Figure J-2

design. The bent-tube design allows for higher rates of burning within the combustion space. Also, superheaters are readily fitted into watertube units.

Industrial watertube units in excess of 250,000 lbs/hr. of steam capacity are rare as are operating pressures of 600 psi. However, the higher pressure units can be found where electrical power is generated. Utility boilers used in large power plants commonly operate with steam produced in the $1,000°$ F range. Utility boilers are almost always of the watertube design and all types of fuels are used with coal being the most common.

❏ Water Treatment/Blowdown

Usually, municipally delivered domestic water is not of sufficient good quality to be used for steam generation feed water and it must be treated prior to use as well as during use. The type of water treatment used is dependent on the delivered water quality, but usually involves controlling boiler tube deposits, boiler tube corrosion, and carryover during use and softening prior to use.

Boiler tubes are particularly susceptible to scale-like deposits when the contacting water is heated to very high temperatures and the equilibrium conditions in the water are upset. (For example, if the water is in contact with a hot surface and the solubility of the contaminant is lower at the higher temperature, then the contaminant will precipitate out on the hot surface - forming scaly deposits.)

The most common contaminants found in boiler scale are calcium and magnesium compounds, iron oxide, silica and alumina. Deposits are a serious problem in boiler tubes in that they degrade heat transfer potential which can ultimately cause a boiler to fail entirely. For example, in high pressure boilers and at the higher heat transfer rates of 900 - 1350° F (480 - 730° C), the steel tubes begin to deteriorate if the circulating water does not conduct the heat away from the metal. When the water tubes become occluded with deposits, these deposits act as insulators and reduce the heat removal rate. This condition produces boiler tube overheating and possible tube blowout causing, at worst, complete system failure. Treatment of the make-up water, steam condensate return, or both, with corrective chemicals is the usual procedure to mitigate scale buildup.

A second major, water-related problem is boiler tube corrosion caused by oxygen attack on steel. This condition exists throughout the boiler system in any place where oxygen is present. Elimination of oxygen in boiler feedwater is the primary way this situation is controlled (e.g., water treatment chemicals referred to as oxygen scavengers, are used to minimize the oxygen levels in the system).

The third consideration in maintaining efficient boiler operations is controlling carryover from the boiler to the steam system. When steam is generated, essentially a pure water vapor is discharged from the boiler, leaving the solids behind. Over a period of time, these solids build up and reach a maximum acceptable concentration, past which efficient boiler operation does not occur. The usual way in which a suitable balance is maintained is by utilization of a boiler blowdown. During blowdown the concentration of chemicals and solids are reduced to an acceptable level. Blowdown can be done two ways: (1) continuous blowdown whereby there is a constant removal of water and solids at a steady rate and (2) intermittent blowdown whereby the solids are allowed to concentrate and are discharged at regular intervals. (Once per day, once per shift or some other selected frequency.) The blowdown valve is opened for a short period of time (usually 1 - 10 minutes) and this effectively reduces the concentration of solids. Many times blowdown is routed to a mud drum and the size of the container can be useful to the inspector in estimating maximum blowdown volume per blowdown occurrence.

An additional aspect of boiler feed water treatment is softening of the actual feedwater prior to additional chemical treatment. The softening is done to eliminate as much of the hardness (Calcium and Magnesium) as possible with a Sodium Zeolite process as shown below:

Softening: $\quad Ca^{+2} + Na_2X \rightarrow CaX + 2 Na^+$

$\quad Mg^{+2} + Na_2X \rightarrow MgX + 2 Na^+$

Regeneration: $\quad 2NaCl + CaX \rightarrow Na_2X = CaCl_2$

$\quad 2NaCl + MgX \rightarrow Na_2X + MgCl_2$

The amount of salt applied to the zeolite bed when regeneration is done is determined by a combination of desired effluent quality and the plant capacity needed between scheduled regeneration cycles. Increased salt equals lower hardness leakage but also results in poorer chemical economy and increased volume of spent brine being discharged to sewer. The protocol of salt load versus regeneration frequency is usually a dictate of production, plant capacity, and ease of scheduling regeneration cycles between peak use times.

❏ Water Use

In utility company steam plants, in addition to the feed water required to make the steam, the most significant water use is for the cooling system. Often, fresh water is used to cool the steam condensers. In the process of contact, the water becomes heated and is subsequently discharged at an elevated temperature. When a power generating facility is in close proximity to an body of water, and discharge permits are available through the state or federal government, the problem of cooling water availability is eliminated. The single most important consideration with this type of discharge is thermal pollution. Opinions vary on this topic, with extreme points of view being held. On the one hand, critics believe that the excessive discharge of heated water destroys the indigenous populations of flora and fauna within the contact zone, seriously impacting the local ecosystem. On the other side, proponents say that deliberately increasing sea water temperature can provide an enhanced breeding and cultivating ground for exotic sea food, such as lobsters, which bring positive economic gains to an area which might have been previously non-productive. The debate continues...

In the smaller "industrial only" applications the cooling water question is more easily resolved with the use of cooling towers. (See section on cooling towers for details.)

The primary concern for the inspector is how to calculate evaporative losses, now involving both steam losses and possibly cooling water losses. The best way to resolve the losses dilemma is to request "make-up volumes for both boilers and cooling towers and allocate portions to evaporative losses and blowdown for both cooling towers and boilers. (This can usually be done using a combination of hard metering data from the contact, engineering specification for the equipment used and/or the use of engineering estimates for cooling towers, boilers, etc.)

❑ **Return Condensate**

Return condensate is that portion of the steam that has cooled sufficiently to revert back to a liquid form. Occasionally, cleanup of this return condensate is required to protect the boiler system. The condensate may have become contaminated with oil, with corrosion products, or from in-leakage of hard water. If the steam plant is exceptionally old, there will most probably be a higher degree of contamination in the return condensate. In some extreme cases it may be more practical and economical to discharge this contaminated return condensate to sewer and only use fresh water for the make-up/feed to the boiler.

❑ Pollutants of Concern

The wastes of concern that could originate from steam plants, other than the cooling water per se, are as follows:

- Hot, concentrated briny water from boiler and evaporator blowdown (See blowdown section)

- Concentrated acid and/or alkaline solutions used for boiler system cleaning operations

- Cooling tower bleed (see cooling water section)

- Scrubber water from stack emission control systems (if applicable)

- Concentrated acid and/or alkaline or briny solutions from regeneration of ion-exchange softeners, demineralization or zeolite softening systems (for feed water supply)

- Miscellaneous testing laboratory wastes

All boilers must be cleaned periodically, both during operations and at regular intervals. High pressure boilers must be cleaned on an average of once per year. Strong acid and alkaline solutions as well as special cleaning agents (hydrochloric acid, acetic acid, potassium bromate, ammonia, detergents and phosphates) are used to descale the entire boiler system. The resultant waste water generated must be disposed of and the sewer is the usual choice. Since this type of cleaning is relatively frequent, many boiler plants have an acid waste problem if no neutralization system or acid-waste treatment is incorporated into their overall operations plan.

Additionally, the demineralization ion-exchange resin requires acid and alkali regeneration every few days. This waste stream presents a second acid-waste disposal problem or pretreatment requirement.

If the plant is of any consequence in size, and the chemical treatment of the fresh water feed is not done by a service (which regenerates the resin containers at their central facility), it seems most appropriate that the facility would need to incorporate some type of acid/alkali waste treatment facility into their operation, or at the least, have an acid waste treatment plan. A suitable plan would incorporate something other than wholesale acid disposal to the sewer).

INSPECTOR'S QUESTIONS—STEAM PLANTS/BOILERS

1. What is the use of the boiler?
    - Hot water?
    - Steam?
    - Other?

2. What type of boiler is in use?
    - Low pressure?
    - High pressure?
    - Firetube boiler?
    - Watertube boiler?
    - Other?

3. Is there any treatment of the boiler feed water?
    - Softening?
    - If yes, what method?
    - Demineralization?
    - If yes, what method?

4. Is the boiler feed water (make-up) metered?
    - If yes, what is the daily average used?
    - Is condensate returned to the system?
    - If yes, what percent?
    - What are the steam losses (gallons per day)?
    - How is that figure calculated?

5. Is there any regeneration of the water softening system in-house?
    - If yes, what is the frequency of regeneration?
    - What is the volume of brine discharged to sewer?
    - Where is the discharge plant located?

6. Is there any regeneration of the demineralization system in-house?
    - Is it on a service?

- If yes, what is the regeneration frequency?
- What is the volume of wastewater discharged?
- Where is the discharge plant located?
- Is there any pretreatment of this wastewater?
- If yes, what is the form of pretreatment?
- If no, does it require pretreatment prior to discharge?
- If no, why does it not require pretreatment prior to discharge?

7. Is there a "blowdown" from the boiler(s)?
    - If yes, what is the frequency?
    - What is the volume?
    - Where is the discharge point located?

8. What water treatment chemicals are used for the boiler?

9. How frequently is preventive maintenance conducted, i.e., cleaning/descaling of the boiler tubes and boiler drum?
    - What other chemicals does the facility use?
    - What acids does the facility use?
    - How does the facility dispose of the wastes generated?
    - What pretreatment is done prior to discharge?

10. Is there any cooling water used?
    - If yes, what type?
    - Single pass?
    - Cooling tower?

11. If single pass...
    - What is it used for?
    - What is the volume (gallons per day)?
    - Where is the discharge point located?
    - Is it commingled with a regulated process stream(s)?
    - (If yes, note this for factoring "diluted" wastestream limits)
    - Is there an NPDES permit for the site?
    - If yes, what is the permit number?

12. If cooling tower...

- How many towers are in use?
- Where are they located?
- What are they used for?
- What is the tonnage of each tower?
- What is the bleed from the tower(s) (gallons per day)?
- Where is the discharge point(s)?
- Are chromates used in the water treatment?
- If yes, where?
- What chemicals are used?
- What are the evaporative losses? (Describe volumes and how derived)

13. Is there adequate spill containment in the processing area(s)?
    - If yes, describe.
    - If no, describe.
    - How is the sewer otherwise protected from leaks or spills in the process area(s) and throughout the facility? Describe in detail.

─────────────── **Section K** ───────────────

Universities, Military Installations and Other Large Institutions

Introduction

Evaluation of large, complex institutions like military bases and college campuses can prove to be a difficult and time consuming experience. The waste streams at these facilities differ from the normal permitted sites in that they can have a very diverse assortment of operations, each producing a waste stream needing evaluation on its own merits. Also, a typical industrial site produces a few relatively high flow, easily identified process streams. This group, on the other hand, is more apt to have several low volume, not-so-easily identified process streams spread out over a very large area. Additional complexity is added if the responsibility for particular sections is divided among operational groups; e.g., boiler plant, physical plant, grounds, building, etc. In the case of the military, division may occur among various tenant facilities as well as in the operational hierarchy. This type of segmentation can cause added delay and confusion during the inspection and permitting process.

Typical operations needing evaluation at a military site or a college or university campus include the following:

- Plant operations - steam plants, boilers, co-generation sites, cooling towers, chillers, refrigeration units, compressors, vacuum systems, etc.

- Maintenance operations - various cleaning operations, steam-cleaning, general upkeep of site

- Metal Finishing - foundry, welding, electroplating, etc.

- Metal Working - machine shops, etc.

- Auto Mechanics - auto hobby shops, automotive repair

- Car Washes - steam cleaning operations

- Carpentry Shops - woodworking shops
- Paint Shops - paint booths
- Photo-printing, graphic arts, silk-screening
- Pottery - ceramics, jewelry making
- Laundries - commercial and/or coin-operated
- Hospitals - clinics, dispensaries
- Laboratories - medical, dental, chemical, biological, etc.
- Dining/Eating Facilities, Commissaries
- Agricultural - landscaping, horticulture
- Hazardous Waste Storage/Treatment Facilities - including radioactive waste storage and/or disposal

For specific wastestream evaluation the inspector can refer to the sections in this manual where these industrial operations are covered in detail. For those operations not found, it is safe to assume that normally only negligible amounts of industrial wastewater are generated.

INSPECTOR'S QUESTIONS—LARGE INSTITUTIONS

1. Is there a map available which details the site and includes all buildings, roads, plus water and sewer lines? If not, the inspector should attempt to draw one, as comprehensive as possible or include important details on whatever is provided.

2. Is there some way to identify all of the water meters associated with this facility? Describe...

3. Is there some way to identify which meters serve which buildings or areas, or are meters looped? Explain... Is there some way to identify all of the sewer laterals, mains and sample points? Explain...

4. Does this site have any active or inactive NPDES permits? Describe the location, the use of the water, and the gallons per day discharged...

5. Is there a master list of the chemicals used which can be provided? (If not, one needs to be procured or developed.)

6. What is the total population at this site?
    - Employees (8-hour)?
    - Live-aboards (dormitories, barracks, etc. - 24-hour)?
    - Single-family residences?

7. Any requirements for steam generation or use?
    -If yes, is make-up water metered?
    -Volume (gallons per day)?
    -Is condensate returned to system?
    -Is there blowdown to sewer? ...Frequency? ...Volume (gallons per day)?
    -What are the steam losses (gallons per day)?

8. Any cooling water in use? Describe cooling towers in use, where located,

etc.

9. Any rectifiers, compressors or similar equipment in use?
   - If yes, do any of them require cooling water?
   - Any one-pass cooling water?
   - If yes, volume (gallons per day)?
   - Where is the discharge point?
   - If water not used for cooling equipment, describe cooling practices.

10. Is any vapor degreasing done?
    - If yes, is the unit water cooled (one-pass), on a cooling tower or refrigerant type?
    - If one-pass, where does it discharge to sewer? (One-pass not usually allowed by sewer ordinance)

11. Is there any metal finishing or metal working done on site? Explain...

12. Are alkaline cleaning or acid cleaning tanks present?
    - If yes, how frequently are these tanks batch discharged to the sewer?
    - What pretreatment occurs prior to discharge?
    - What are the individual volumes of all process tanks in facility - plating, acid, etc.?
    - How is bottom sludge handled?

13. What are the individual volumes and locations of the process tanks?

14. How are spent or contaminated processing tanks handled?
    - If hauled, who hauls it? (Can check manifests.)

15. Is there any routine washdown of the work area(s)?
    - Where is the discharge point?

16. Are there any floor drains in the process area?
    - Does their flow pass through a common interceptor point?
    - If no, where does it discharge?

17. Any auto mechanics, maintenance, washing, etc.? Describe...

18. Any photo, printing, graphic arts? Describe and include sample points if applicable...

19. Any laundries? Describe...

20. Any hospitals, clinics, etc.? Describe...

21. Any laboratories? Describe...

22. Any woodworking, carpentry, painting, paint booths or water curtains? Describe...

23. Identify all eating or dining facilities and provide number of meals served per day.
- Is there an automatic dish washer present?
- Hours of operation?
- Gallons per day discharged to sewer?
- Is there a grease trap or interceptor present? Describe...

24. Describe the routine handling or herbicide and pesticide residuals and empty containers, if applicable.

25. Describe the hazardous waste storage and treatment if applicable - including radioactive waste if present.

26. How is the routine disposal of hazardous wastes handled?
- Who is the waste hauler?
- What is the average volume of material disposed to landfill, recycled, etc.?
- Where are the records kept? (Can ask to see manifests.)

27. Do all process streams flow to a common interceptor point?
- If yes, where is that point?
- If no, where are the various points?

28. Is there any pretreatment of waste water?
- If yes, describe the various aspects in place and where located.

29. Is there any water reuse in the facility?
    - If yes, where and what is it?

30. What solvents are used at this facility?
    - Where are they used? What unit operations?
    - What is the method of application? Dip tank, rag applied, or other?
    - Is there any discharge of solvents to sewer in any form? From the unit operation(s)? Final triple rinsing of containers prior to disposal? Explain...
    - If no solvents are discharged to sewer, explain what procedures are in place to prevent discharge. Solvent Management Plan (SMP)?
    - When was SMP submitted? Explain...

31. Is there adequate spill containment in the processing area(s)?
    - If yes, describe.
    - If no, describe.
    - How is the sewer otherwise protected from leaks or spills in the process area(s) and throughout the facility? Describe in detail.

―――― **Section L** ――――

METAL FINISHING/ELECTROPLATING

INTRODUCTION

Among facilities that discharge potential pollutants into the nation's waterways, metal finishing facilities - and particularly electroplaters - account for the majority of metals, cyanides, and toxic organic wastes. Most of these constituents threaten both aquatic life and human health. The Environmental Protection Agency (EPA) has set stringent limits for these chemicals in surface waters. Typically, concentrations of these materials in untreated discharges from metal finishers are several times greater than allowed by the regulations. Even when diluted by a large volume of receiving water, untreated metal finishing discharges can severely degrade water quality.

The metal finishing industry as a generic group is subject to categorical standards (40 CFR Parts 413 and 433) as established by the EPA. The goal of these regulations is to <u>reduce the contaminants in metal finishing discharges to levels that are environmentally acceptable while remaining technically feasible and affordable for the industry.</u>

The EPA has developed standards for facilities that discharge to surface waters (known as direct dischargers) and for facilities that discharge to publicly owned treatment works (POTW) (known as indirect dischargers). <u>This section deals primarily with Categorical Pretreatment Standards and their applicability to indirect dischargers.</u> The General Pretreatment Regulations (40 CFR Part 403) specify that POTWs or local sewerage agencies will be responsible for administering local pretreatment programs and implementing and enforcing the categorical pretreatment standards. This section looks at these regulations and how they affect local programs and provides summary information in several key areas.

- ❏ **General Pretreatment Regulations, Categorical Pretreatment Standards and Local Limits**

- ❏ **An overview of electroplating and metal finishing**

- ❏ **How to distinguish a metal finisher from an electroplater**

- **An evaluation of Industry Operations**

- **Process Operations/Overview**
    - **Electroplating/Inspector's Questions**
    - **Electroless Plating/Inspector's Questions**
    - **Anodizing/Inspector's Questions**
    - **Conversion Coating/Inspector's Questions**
    - **Etching (Chemical Milling)/Inspector's Questions**
    - **Printed Circuit Boards/Inspector's Questions**

- **Additional 40 Metal Finishing Process Operations**

- **Sources of Water Pollution**

- **Waste Water Technologies for Water Conservation and Waste Minimization**

- **Standard Waste Treatment Technologies**

- **Combined Waste Stream Formula**

- **Compliance Dates**

❏ General Pretreatment Regulations, Categorical Pretreatment Standards And Local Limits

• General Pretreatment Regulations

The National Pretreatment Program established an overall strategy for controlling the introduction of nondomestic wastes into publicly owned treatment works (POTWs) in accordance with the overall objectives of the Clean Water Act. Sections 307(b) and (c) of the Act authorized the Environmental Protection Agency (EPA) to develop National Pretreatment Standards for new and existing dischargers to POTWs. The Act also made these pretreatment standards enforceable against dischargers to POTWs.

The General Pretreatment Regulations (40 CFR Part 403) further established administrative mechanisms requiring nearly 1,700 POTWs to develop local pretreatment programs to enforce the general discharge prohibitions and the specific Categorical Pretreatment Standards. These Categorical Pretreatment Standards were designed to prevent the discharge of pollutants which could pass through, interfere with, or were otherwise incompatible with the operations of the POTWs. The standards imposed are technology-based and mandate the removal of toxic pollutants. They also contain specific numerical limitations based on evaluations of specific technologies for the particular industrial categories in question.

As a result of a settlement agreement, the EPA was required to develop Categorical Pretreatment Standards for 34 industrial categories with a primary emphasis on 65 classes of toxic pollutants.

❑ Categorical Pretreatment Standards - 40 CFR Part 413 - Electroplater Standards

Electroplater Standards established in 1979 set specific numerical limitations for seven groups of dischargers. Several industry groups filed petitions to review the limits set by EPA and subsequently on 28 January 1981 a settlement was reached by EPA and the industry groups. These changes were incorporated into the 1981 amendments to the Electroplating Standards 40 CFR Part 413. One of the more significant changes to these Electroplating Standards was the development of additional pretreatment standards called "Metal Finishing" Standards, which now regulate processes previously regulated under electroplating standards. The EPA stated that in light of the potentially severe economic impact of these regulations on the job shop and independent printed circuit board manufacturers segment of the industry, EPA would not impose the more stringent pretreatment standards on that segment of the industry until later. (See page L-5-b.)

The job shop electroplaters and independent printed circuit board shops who had been in operation prior to 31 August 1982 have remained, and will always remain, regulated under the 40 CFR Part 413 Electroplater Limitations. It was EPA's anticipation that the less stringent limits would pass out of use, or be phased out, due to natural attrition. Now, when new job shop electroplaters and independent printed circuit board manufacturers petition for discharge permits, they are given the more stringent PSNS limits. (See page L-5-b.)

ELECTROPLATING LIMITATIONS
40 CFR Part 413

Pollutant	<10,000 gpd of regulated process flow	>10,000 gpd of regulated process flow
Cadmium (T)	1.2 mg/l	1.2
Chromium (T)		7.0
Copper (T)		4.5
Lead (T)	0.6	0.6
Nickel		
Silver (T)[1]		1.2
Zinc (T)		4.2
Cyanide (T)		1.9
Cyanide (A)	5.0	
Total metals [sum concentration of (T) chromium, (T) copper, (T) lead, (T) nickel]		10.5
Total Toxic Organics	4.57 mg/l	2.13 mg/l

[1] Applies to precious metal electroplating only

These limits apply to all job shop electroplaters and independent printed circuit board manufacturers who were in operation prior to August 31, 1982.

❑ Categorical Pretreatment Standards 40 CFR Part 433 - Metal Finishing Standards

In accordance with the Agency's plan, EPA promulgated the Metal Finishing Categorical Pretreatment Standards on 15 July 1983 as 40 CFR Part 433. The effect of this new standard was to create a new category - Metal Finishing - which most electroplaters would have to comply with following their compliance with the Electroplating Standard. These new limits would apply uniformly to dischargers from electroplating and other metal finishing operations. This would satisfy industry requests for equivalent limits for process lines often found together and would greatly reduce the need for the combined waste stream formula. (See page L-76.)

Once the compliance date, 15 February 1986, for the Metal Finishing Standards had been reached, all firms conducting one or more of the six basic operations of the Electroplating Category (See Table L-5 on page L-13-a) were required to be in compliance with the Metal Finishing Pretreatment Standards. The only exceptions being job shop electroplaters and independent printed circuit board manufacturers (those in operation prior to 31 August 1982) who would continue to be regulated under Part 413 of the Electroplating Pretreatment Standards. They are and will be forever exempt from Part 433 of the Metal Finishing Standards. (See Tables L-1, L-2, and L-3 on pages L-6-a, L-6-b, and L-6-c.)

Table L-1

METAL FINISHING PRETREATMENT STANDARDS
EXISTING SOURCE (PSES) LIMITATIONS

POLLUTANT OR POLLUTANT PARAMETER	DAILY MAXIMUM	MAXIMUM MONTHLY AVERAGE
Cadmium	0.69	0.26
Chromium, total	2.77	1.71
Copper	3.38	2.07
Lead	0.69	0.43
Nickel	3.98	2.38
Silver	0.43	0.24
Zinc	2.61	1.48
Cyanide, total	1.20	0.65
TTO (Flow < 10,000 gpd)	4.57	
TTO	2.13	
Alternative to total cyanide:		
Cyanide, amenable to chlorination	0.86	0.32

PSES Metal Finishing Standards: Apply to electroplating and/or metal finishing facilities which existed prior to August 31, 1982 and are not job shop electroplators or printed circuit board manufacturers.

Table L-2

METAL FINISHING PRETREATMENT STANDARDS
NEW SOURCES (PSNS) LIMITATIONS

(FOR ANY NEW SOURCE)

POLLUTANT OR POLLUTANT PARAMETER	DAILY MAXIMUM	MAXIMUM MONTHLY AVERAGE
	Milligrams per liter (mg/l)	
Cadmium	0.11	0.07
Chromium, total	2.77	1.71
Copper	3.38	2.07
Lead	0.69	0.43
Nickel	3.98	2.38
Silver	0.43	0.24
Zinc	2.61	1.48
Cyanide, total	1.20	0.65
TTO	2.13	
Alternative to total cyanide:		
Cyanide, amenable to chlorination	0.86	0.32

PSNS Metal Finishing Standards: Apply to electroplating and metal finishing facilities which began their operations after August 31, 1982, the date of the proposed regulation.

Table L-3

METAL FINISHING PRETREATMENT STANDARDS
NEW SOURCES (PSNS) LIMITATIONS

FOR INDUSTRIAL FACILITIES WITH CYANIDE TREATMENT

POLLUTANT OR POLLUTANT PARAMETER	DAILY MAXIMUM	MAXIMUM MONTHLY AVERAGE
	Milligrams per liter (mg/l)	
Cyanide (A)	0.86	0.32

❏ Local Limits

In addition to federally mandated limits, electroplaters and metal finishers are also subject to local limits. Local limits are <u>always applicable</u> for <u>all dischargers</u> and are only changed in the instance of a higher authority mandating more stringent limits. (This is the case with categorical limits, whereby a local limit may become more stringent because of a federal intervention.)

Also, in most instances local limits are regulating conventional pollutants as well as toxic pollutants. The conventional pollutants are regulated because of problems that could be caused by excessive amounts entering the sewer system. An example of a local limit is 500 mg/l of grease and oil. The local authority has deemed that grease and oil in excess of 500 mg/l from any commercial or industrial discharger can cause problems for the system as a whole.

❏ An Overview of Electroplating and Metal Finishing

There are 46 distinct metal finishing operations, of which six are regulated separately as electroplaters (see Table L-5, Page L-13-a). Metal finishing operations are regulated by the Environmental Protection Agency (EPA) under the Metal Finishing Categorical Pretreatment Standards (enacted 15 July 1983) known as 40 CFR Part 433. The Electroplating Pretreatment Standards are found in 40 CFR Part 413 and were first enacted on 7 September 1979, revised 28 January 1981 and now regulate only independent printed circuit board manufacturers and job shop electroplaters in operation prior to August 31, 1982.

Industries considered to be "electroplaters" are those performing electroplating, electroless plating, anodizing, coatings, chemical etching/milling or printed circuit board manufacturing <u>and</u> who initiated their discharge <u>prior to 31 August 1982.</u>

All new sources, regardless of category type, who initiated discharge after 31 August 1982 must now comply with the Metal Finishing Standards 40 CFR Part 433 (PSNS) rather than 40 CFR Part 413 ELECTROPLATING STANDARDS or even 40 CFR 433 (PSES) METAL FINISHING STANDARDS.

What this means is that those industries which previously would have been regulated under the Electroplating Pretreatment Standards 40 CFR 413, must now instead comply with the more stringent Metal Finishing Standards 40 CFR 433. The two notable exceptions being existing independent printed circuit board manufacturers and job shop electroplaters who will continue to be regulated by 40 CFR Part 413. These shops will still be permitted and regulated as "electroplaters", but it is the EPA's expectation that such facilities will eventually phase out over time. Gradually, the entire industry base will come to be regulated only under Metal Finishing Standards 40 CFR Part 433 (PSES) or (PSNS).

❑ How To Distinguish a Metal Finisher From An Electroplater

<u>Existing Facilities versus New Facilities:</u>

Determination of type of facility status is based on the date when <u>discharge</u> began. Existing facilities are those which began discharge prior to August 31, 1982. New facilities are so classified if they began discharge after August 31, 1982.

<u>Electroplater Standards versus Metal Finisher Standards (PSES and PSNS):</u>

Electroplating (PSES*)

1. Began discharge prior to August 31, 1982. (*Note: only PSES Standards exist for electroplaters but there are two sets of limits; <10,000 gpd and >10,000 gpd.) (See page L-3)

2. Own <50% of materials undergoing metal finishing, i.e., job shop electroplater or independent printed circuit board manufacturer.

Metal Finishing (PSES)

1. Began discharge prior to August 31, 1982

2. Perform at least one of the six qualifying operations (electroplating, electroless plating, anodizing, conversion coatings, chemical milling/etching and/or printed circuit board manufacture)

3. Are not subject to electroplating PSES, i.e., owns >50% of the materials undergoing metal finishing, etc.

* Note: MOVEMENT TO ONE SET OF STANDARDS

Currently, there are two sets of standards regulating the metal finishing industry: those regulating metal finishing pretreatment (40 CFR Part 433) and those regulating electroplating pretreatment (40 CFR Part 413). Historically, only one set of standards existed for the industries when they were categorized as electroplaters. In 1982 these standards were revised and metal finishing pretreatment standards were added. Industries and businesses now regulated as electroplaters will be phased out over time, and new businesses (since August 31, 1982) have been given permits with the more stringent metal finishing limits.

Metal Finishing (PSNS)

    1. Began discharge after August 31, 1982

    2. Performs at least one of the six qualifying operations above

❑ **Evaluation of Industry Operations**

Initially, an evaluation of the Industry's operation must be made to identify certain key processes and collect operational data that will be needed throughout the inspection:

- Determine the water flow through the process
    - Identify all inputs, outputs and losses
    - Identify all potentially harmful constituents in waste stream

- Quantify or estimate the flows of all regulated process wastestreams, and all other wastestreams or dilution streams that are mixed with the regulated waste stream

- Determine if water flow varies and why: because of production rate, work shift, batch discharge

- Identify existing sample locations and determine new locations if existing ones are inadequate, inappropriate or non-existent

❑ PROCESS OPERATIONS/OVERVIEW

90% of all water used in plating operations is used for rinse water. Such water will contain wastes from the various chemical process tanks in the process line. This section will describe the electroplating and metal finishing process operations, principle work steps found in these operations, provide specific information about each of the operations and also furnish "Questions for Inspectors" - a list of questions to guide the site inspector.

Electroplating Process Operations/40 CFR Part 413

Electroplating is a process for applying a thin metal coating such as zinc, copper, nickel, chromium, etc. to the surface of metal parts. The metal parts are usually made of iron, steel, brass or aluminum. The coatings serve to protect the metal from corrosion, to build up the surface thickness, or to decorate the piece. Many commonly used items are electroplated. For example, automobile parts are often chrome and nickel plated; printed circuit boards are copper plated; and jewelry can be plated with precious metals such as gold or silver.

Not all the processes regulated by the Electroplating Federal Categorical Standards are true electroplating. Listed below are the six processes regulated by 40 CFR Part 413. They include:

- Electroplating
- Electroless plating
- Anodizing
- Coating
- Chemical etching and milling
- Printed Circuit Board manufacturing

Metal Finishing Process Operations/40 CFR Part 433

The standards for metal finishing regulate 46 process operations. These include the six previously identified as being regulated by the Electroplating regulations and 40 additional processes. (See Table L-5 on page L-13-a.)

NOTE: The Metal Finishing Standards for the 40 different processes apply only if one of the six electroplating processes are also present at the facility.

PROCESSES REGULATED BY METAL FINISHING STANDARDS

1 Electroplating
2 Electroless Plating
3 Anodizing
4 Conversion Coating
5 Etching (Chemical Milling)
6 Printed Circuit Board Manufacturing
7 Cleaning
8 Machining
9 Grinding
10 Polishing
11 Barrel Finishing (Tumbling)
12 Burnishing
13 Impact Deformation
14 Pressure Deformation
15 Shearing
16 Heat Treating
17 Thermal Cutting
18 Welding
19 Brazing
20 Soldering
21 Flame Spraying
22 Sand Blasting
23 Other Abrasive Jet Machining
24 Electric Discharge Machining
25 Electromechanical Machining
26 Electron Beam Machining
27 Laser Beam Machining
28 Plasma Arc Machining
29 Ultrasonic Machining
30 Sintering
31 Laminating
32 Hot Dip Coating
33 Sputtering
34 Vapor Plating
35 Thermal Infusion
36 Salt Bath Descaling
37 Solvent Degreasing
38 Paint Stripping
39 Painting
40 Electrostatic Painting
41 Electropainting
42 Vacuum Metalizing
43 Assembly
44 Calibration
45 Testing
46 Mechanical Plating

* First six processes are also regulated by the Electroplating Standards

Table L-5

Principle Work Steps

Characteristic to most electroplating and metal finishing operations are three principal steps; surface preparation, plating, and a post-treatment process. (See Figure L-1, page L-14-a.) The principle work steps (and their sub-steps) may occur singly or in combination at a particular process operation. The site must include one of the six electroplating process operations in order to be regulated. However, these three principle work steps are also regulated by the following considerations:

- Surface preparation is never regulated in isolation (i.e., when it is the only work step at the shop)

- Plating is always regulated

- Post treatment is regulated if it involves chromating, passivating, phosphating, or some other regulated coatings operation.

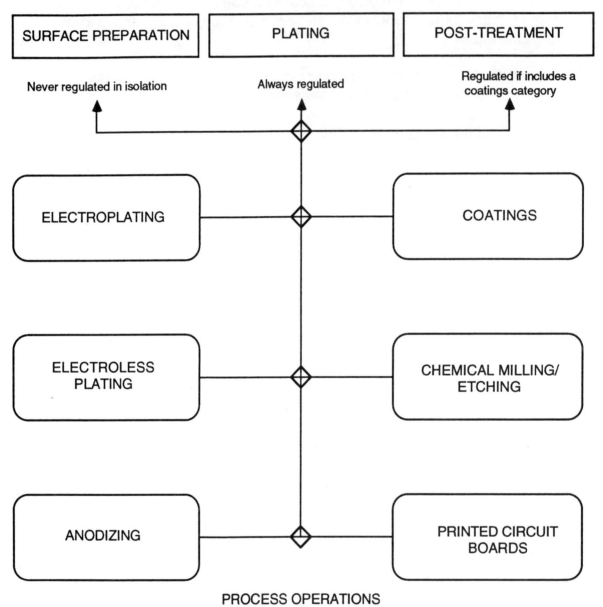

PROCESS OPERATIONS
Figure L-1

L-14-a

- **Surface Preparation**

    Surface preparation involves steps to clean the part before it is plated. Cleaning is usually accomplished by placing the work piece in a tank containing a solvent or alkaline solution, and then in an acid dip to remove corrosion. Both the alkaline and the acid dip are followed by rinsing in running water.

    - Vapor Degreasers: organic solvents (non-flammable) such as 1-1-1 trichlorethane are used in closed system equipment: One-pass cooling water is frequently associated with closed-loop systems. Wastes and sludges must normally be specially treated and do not usually form part of waste streams.

    - Di-phase Cleaning: A two layer system of water soluble and water insoluble organic solvents. As in vapor degreasing a solvent type waste is produced.

    - Alkaline Cleaning: Alkaline cleaners usually contain: sodium hydroxide, sodium carbonate, sodium metasilicate, sodium phosphate (di- or tri-sodium), sodium silicate, sodium tetra phosphate, and wetting agents. Concentrations are usually low when found in associated rinse waters. Tanks may or may not be periodically dumped and cleaned.

    - Acid Cleaning: Acids used are hydrochloric acid (most common), sulfuric acid, nitric acid, chromic acid, fluoboric acid, and phosphoric acid. Concentrations in rinse waters are often quite significant. Baths have a short life and are replaced or replenished frequently. Sometimes they can be discharged to the sewer if neutralized and handled appropriately.

    - Emulsion Cleaning: Common organic solvents are used in emulsions of various phase configurations. Solvent waste produced.

    - Ultrasonic Cleaning: Mechanical method. High effectiveness but also high cost. No waste produced.

- **Descaling**

    - For removal of difficult oxides, particularly in relation to the plating of stainless steel and other corrosion resistant materials, a molten salt bath followed by water quenching and acid dipping is used. Specialized oxidizing, reducing, and electrolytic baths may also be employed.

- **Plating Application**

In the second step, a metallic coating is applied from a solution containing the plating metal in dissolved form, often in association with other chemicals. The part to be plated is placed in the solution which may also be charged with electricity to attract the dissolved metal to its surface much like a magnet attracts iron filings. Plating is followed by a rinsing step to flush the process solution from the work piece. (See pages L-19 and L-23, "Electroplating" and "Electroless Plating", and Figure L-2, page L-18-a.)

- **Post Treatment**

    Often plating steps are followed by some type of post-treatment of the work piece to either decorate it or to add corrosive resistance. These treatments include many types of coatings, but especially conversion coatings. Included are chromating, passivating, phosphating, and the various metal coloring and sealing procedures. However, some electroplating operations are complete after the plating step and do not require post-treatment.

    When post-treatment operations are present, they are normally associated with rinsing steps and present similar pollutant concerns as does the plating operation. These processes typically contribute the following pollutants to the waste stream:

    - Chromium, trivalent and hexavalent
    - Zinc
    - Nickel
    - Cyanide
    - Low pH wastes

NOTE: Occasionally the configuration of a process line will be difficult to analyze because of space constraints, work flow requirements, poor tank layout, inadequate or inappropriate identification of applicable work steps in a process line, or the absence of the third step. For these reasons, the three principal work steps of typical electroplating or metal finishing operations can be difficult to recognize.

OVERVIEW OF THE ELECTROPLATING/METAL FINISHING PROCESS

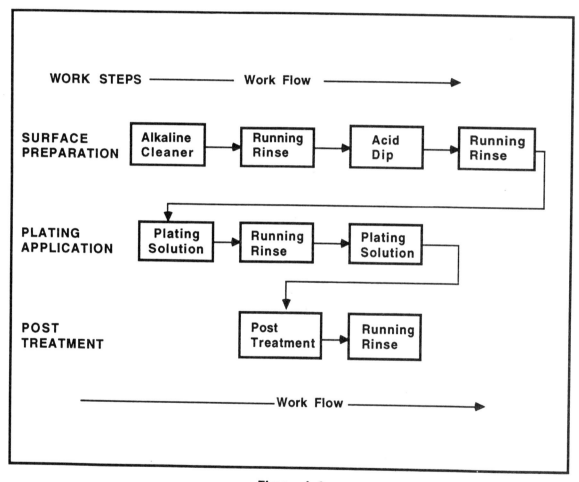

Figure L-2

L-18-a

THE ELECTROPLATING CATEGORY - THE SIX PROCESS OPERATIONS REGULATED BY 40 CFR PART 413

❑ Electroplating

In the electroplating process metal ions in solution are reduced on the cathodic (negatively charged) surface of the workpiece being plated. The ions in solution are replenished by the dissolution of metal from the anode (positively charged) which consists of small pieces of metal to be plated. Occasionally (chromium plating for example) metal ions are replenished from solutions of metal salts. In these cases the anode is usually some inert material like lead.

The above process results in the production of a thin coating of one metal upon another by electrodeposition. Ferrous or non-ferrous materials may be coated by a variety of common (copper, nickel, lead, chromium, brass, bronze, zinc, tin, cadmium, iron, aluminum or combinations thereof) or precious (gold, silver, platinum, osmium, iridium, palladium, rhodium, indium, ruthenium, or combinations thereof) metals.

The electroplating baths contain metal salts, alkalies and other bath control compounds in addition to plating metals such as copper, nickel, silver or lead. Many plating solutions contain metallic, organo-metallic and organic additives to induce grain refining, leveling of the plating surface and deposit brightening.

INSPECTOR'S QUESTIONS—ELECTROPLATING

1. When did electroplating operations begin at this facility? Day/month/year?

2. Is this facility an independent job shop? Or are the materials being processed owned by the company? Describe if startup occurred prior to August 31, 1982.

3. What metals are being plated?

4. What other metals are in use, in solution or otherwise present in process tanks?

5. What metals are complexed with cyanide?

6. Is any vapor degreasing done?
    - If yes, is the unit water cooled (one-pass), on a cooling tower or refrigerant type?
    - If one-pass, where does it discharge to sewer? What is the flow rate? How many hours per day? (One-pass not usually allowed by local sewer ordinance)
    - What type of solvent is used in the vapor degreaser?
    - How is the spent or contaminated solvent handled? Disposed of? Describe...

7. Are alkaline cleaning or acid cleaning tanks present?
    - If yes, how frequently are these tanks batch discharged to the sewer?
    - What pretreatment occurs prior to discharge?
    - What are the individual volumes of all process tanks in facility - plating, acid, etc.?
    - How is bottom sludge handled?

8. How are spent or contaminated plating baths or other tanks handled for disposal?
    - If hauled, who hauls it? (Can ask to see manifests.)

9. Are any tanks heated?
    - If yes, which ones?

10. What are the individual volumes and locations of the process tanks?

11. Any rectifiers, compressors or similar equipment in use?
    - If yes, do any of them require cooling water?
    - Any one-pass cooling water?
    - If yes, volume (gallons per day)?
    - Where is the discharge point?

12. Is there any reverse osmosis or de-ionized water production?
    - If yes, is it on a service? Who maintains the equipment?
    - What, if any, pretreatment is there for the systems? Describe...
    - Is water kept in a holding tank and drawn off as needed?
    - Or is it produced on an as-needed basis?
    - Is the water metered in? Out?
    - What is the reject ratio for the reverse osmosis system? I.e., what percentage of water goes back to the sewer as reject and what percentage is used as process water?
    - Where is the sample point for the wastewater discharge?
    - Does regulated wastewater from the electroplating operation also flow to this location?

13. Is there routine washdown of the work area(s)?
    - Where is the discharge point?

14. Are there any floor drains in the process area?
    - Does their flow pass through a common interceptor point?
    - If no, where does it discharge?

15. Do all process streams flow to a common interceptor point?
    - If yes, where is that point?
    - If no, where are the various points?
    - Is one-pass or brine reject water co-mingling with categorically regulated wastewater at the sample point(s)?
    - (See Combined waste stream Formula)

16. Is there any pretreatment of the waste water?
    - If yes, describe the various aspects in place and where they are located
    - If no, make recommendations or appropriate comments

17. How is routine disposal of hazardous wastes handled?
    - Who is the waste hauler?
    - What is the average volume of material disposed to landfill, recycled, etc.?
    - Where are the records kept? (Can ask to see manifests)

18. Is there any water reuse in the facility?
    - If yes, where and what is it? Describe...

19. What types of rinse tanks, rinsing procedures or configurations are in use at this facility? Explain or describe...
    - Running rinses
    - Still rinses
    - Countercurrent rinses
    - Spray rinses
    - Fog rinses
    - Other

20. What solvents are used at this facility?
    - Where are they used? What unit operations?
    - What is the method of application? Dip tank, rag applied, or other?
    - Is there any discharge of solvents to sewer in any form? From the unit operation(s)? Final triple rinsing of containers prior to disposal? Explain...
    - If no solvents are discharged to sewer, explain what procedures are in place to prevent discharge. Solvent Management Plan (SMP)?
    - When was SMP submitted? Explain...

21. Is there adequate spill containment in the processing area(s)?
    - If yes, describe.
    - If no, describe.
    - How is the sewer otherwise protected from leaks or spills in the process area(s) and throughout the facility? Describe in detail.

❑ Electroless Plating

Electroless plating as an operation is not usually found in isolation, as some other plating steps are, but rather is done is combination with another electroplating operation, such as those found in printed circuit board shops. The electroless plating step is a chemical reduction process which depends on the reduction of a metallic ion in an aqueous solution containing both a reducing agent and a catalyst. Metal is deposited without the use of an external electric current as is the case in true electroplating. A uniform coating is produced on all surfaces regardless of workpiece geometry. The basic constituents in an electroless plating solution are:

- A source of metal, usually a salt
- A reducer to reduce the metal to its base state
- A chelating agent to hold the metal in solution (for a list of common chelating agents, see Table L-6)
- Various buffers and other chemicals designed to maintain bath stability and increase bath life
- Surface preparation followed by the application of some type of a catalyst prior to the electroless step

- Steps in electroless plating on plastic:

    -Cleaning; acid or alkaline bath
    -Roughening or etching; mechanical roughening and/or etching with acids (particularly chromic acid)
    -Catalyst application; deposition of a thin layer of palladium usually in association with tin

- Steps in electroless plating on metal:

    -Cleaning; vapor degreasing or traditional baths
    -Smoothing; mechanical honing followed by acid dipping and alkaline cleaning
    -Catalyst application; usually a flash deposit or preplating with nickel or copper

Table L-6

COMMON CHELATING AGENTS

HYDROXY ACIDS	Glycolic Acid Gluconic Acid Citric Acid Tartrates
AMINES	tetren trien TBED DPDEED Thiourea TPA PPDT TPED TEA
AMINOCARBOXYLIC ACIDS	EGTA CDTA DHEG HEIDA DTPA DTA HEDTA EDTA

INSPECTOR'S QUESTIONS—ELECTROLESS PLATING

1. When did electroplating operations begin at this facility? Day/month/year?

2. Is this facility an independent job shop? Or are the materials being processed owned by the company? Explain...

3. What metals are being plated? (Nickel and copper are the most common.)

4. What chelating agents are used to hold the metal(s) in solution?

5. How are spent plating baths handled?

6. Are there any other metal bearing solutions present in the process tanks or from other chemical solutions used or other operations at the facility?

7. Are there any cyanide bearing solutions or waste streams present?
    - If yes, from which process(es)?
    - Where does it (they) discharge?

8. Is any vapor degreasing done?
    - If yes, is the unit water cooled (one-pass), on a cooling tower or refrigerant type?
    - If one-pass, where does it discharge to sewer? (One-pass not usually allowed by local sewer ordinance.)
    - What type of solvent is used in the vapor degreaser?
    - How is the spent or contaminated solvent handled? Disposed of? Describe...

9. What are the individual volumes and locations of the process tanks?

10. Are alkaline cleaning or acid cleaning tanks present?
    - If yes, how frequently are these tanks batch discharged to the sewer?
    - What pretreatment occurs prior to discharge?

- What are the individual volumes of all process tanks in facility - plating, acid, etc.?
- How is bottom sludge handled?

11. How are spent or contaminated processing tanks handled?
    - If hauled, who hauls it? (Can check manifests.)

12. Are any tanks heated?
    - If yes, which ones?

13. Any rectifiers, compressors or similar equipment in use?
    - If yes, do any of them require cooling water?
    - Any one-pass cooling water?
    - If yes, volume (gallons per day)?
    - Where is the discharge point?
    - If water is not used for cooling equipment, describe cooling practices.

14. Is there any routine washdown of the work area(s)?
    - Where is the discharge point?

15. Are there any floor drains in the process area?
    - Does their flow pass through a common interceptor point?
    - If no, where does it discharge?

16. Do all process streams flow to a common interceptor point?
    - If yes, where is that point?
    - If no, where are the various points?
    - Is there any co-mingling of unregulated wastewater with categorically regulated wastewater? Explain...

17. Is there any pretreatment of waste water?
    - If yes, describe the various aspects in place and where located.

18. How is the routine disposal of hazardous wastes handled?
    - Who is the waste hauler?
    - What is the average volume of material disposed to landfill, recycled, etc.?
    - Where are the records kept? (Can ask to see manifests.)

19. Is there any water reuse in the facility?
    - If yes, where and what is it?

20. Is there any reverse osmosis or de-ionized water production at this facility?
    - If yes, is it on a service? Who maintains the equipment?
    - What, if any, pretreatment is there for the systems? Describe...
    - If reverse osmosis is produced, what is the reject ratio...i.e., what percentage of water goes back to sewer and what percentage is used as process water?
    - Is water kept in a holding tank and drawn off as needed?
    - Or is it produced on an as-needed basis?
    - Is the water metered in? Out?

21. What solvents are used at this facility?
    - Where are they used? What unit operations?
    - What is the method of application? Dip tank, rag applied, or other?
    - Is there any discharge of solvents to sewer in any form? From the unit operation(s)? Final triple rinsing of containers prior to disposal? Describe...
    - If no solvents are discharged to sewer, explain what procedures are in place to prevent discharge.
    - Solvent Management Plan (SMP)?
    - When was SMP submitted?

22. What types of rinse tanks, rinsing procedures or configurations are in use at this facility? Explain or describe...
    - Running rinses
    - Still rinses
    - Countercurrent rinses
    - Spray rinses
    - Fog rinses
    - Other

23. Is there adequate spill containment in the processing area(s)?
    - If yes, describe.
    - If no, describe.
    - How is the sewer otherwise protected from leaks or spills in the process area(s) and throughout the facility? Describe in detail.

❏ ANODIZING

Anodizing is an electrolytic oxidation process which converts the surface of a metal to an insoluble oxide. Such oxides provide corrosion protection, decoration, and a basis for other coating processes. The most commonly anodized metal is aluminum followed by zinc, magnesium and titanium.

The metal is prepared by soak cleaning with an alkaline cleaner or a phosphoric acid solution. Cleaning etches the metal slightly to insure an active surface for anodizing. In the case of aluminum the metal is then immersed in sulfuric and chromic acid solutions and rinsed. Magnesium anodizing solutions consist of mixtures of fluoride, phosphate and chromic acids or of potassium hydroxide, aluminum hydroxide, and potassium fluoride. Chromic acid anodic coatings are considered to be more protective than sulfuric acid coatings and are used if a complete rinsing of the part cannot be achieved.

Anodizing wastewater typically contains the basis material and either chromic or sulfuric acid. When dyeing or anodized coatings occurs, the wastewaters will contain chromium or other metals from the dye. Other potential pollutants include nickel acetate (used to seal anodic coatings) or other complexes and metals from dyes and sealers.

INSPECTOR'S QUESTIONS—ANODIZING

1. When did anodizing operations begin at this facility? Day/month/year?

2. Is this facility an independent job shop? Or are the materials being processed owned by the company? Describe if startup occurred prior to August 31, 1982.

3. What type of metal is being anodized? (Most common is aluminum, then zinc, magnesium and titanium.)

4. For what purpose will the anodized metal parts or pieces be used? (Usually aeronautical and/or electronics.)

5. Is there any metal coloring associated with your anodizing operation?
    - Any organic dyes?
    - Any metallic component in the dye formulation?
    - If yes, what type of discharge?
    - Where is the discharge point?
    - Does it eventually flow to a common point?

6. What preparatory cleaning step is present?
    - Alkaline soak?
    - Acid cleaner? (Usually done to etch the piece slightly for better coating adhesion.)
    - Any chromic compounds in alkaline, acid, or deoxydizing baths?

7. What type of acid solution is used to anodize the work piece? (For aluminum, usually sulfuric and chromic acids.)

8. What other metal(s) bearing solutions are present in process tanks for from other chemical solutions or operations at this facility?

9. Is any vapor degreasing done?
    - If yes, is the unit water cooled (one-pass), on a cooling tower or refrigerant type?

- If one-pass, where does it discharge to sewer? (Usually not allowed by local sewer ordinance.) What is the flow rate? How many hours per day?
- What type of solvent is used in the vapor degreaser?
- How is the spent or contaminated solvent handled? Disposed of? Describe...

10. What are the individual volumes and locations of the process and rinse tanks?

11. Are alkaline cleaning or acid cleaning tanks present?
    - If yes, how frequently are these tanks batch discharged to the sewer?
    - What pretreatment occurs prior to discharge?
    - What are the individual volumes of all process tanks in facility - plating, acid, etc.?
    - How is bottom sludge handled?

12. How are spent or contaminated processing tanks handled?
    - If hauled, who hauls it? (Can check manifests.)

13. Are any tanks heated?
    - If yes, which ones?

14. Any rectifiers, compressors or similar equipment in use?
    - If yes, do any of them require cooling water?
    - Any one-pass cooling water?
    - If yes, volume (gallons per day)?
    - Where is the discharge point?
    - If water is not used for cooling equipment, describe the cooling practices.

15. Is there any routine washdown of the work area(s)?
    - Where is the discharge point?

16. Are there any floor drains in the process area?
    - Does their flow pass through a common sample point?
    - If no, where does it discharge?

17. Do all process streams flow to a common sample point?
    - If yes, where is that point?

- If no, where are the various points?
- Is one-pass or brine reject water co-mingling with categorically regulated wastewater at the sample point(s)?

18. Is there any pretreatment of waste water?
    - If yes, describe the various aspects in place and where located.

19. How is the disposal of hazardous wastes handled?
    - Who is the waste hauler?
    - What is the average volume of material disposed to landfill, recycled, etc.?
    - Where are the records kept? (Can ask to see manifests.)

20. Is there any water reuse in the plant?
    - If yes, where and what is it?

21. Is there any reverse osmosis or de-ionized water production at this facility?
    - If yes, is it on a service? Who maintains the equipment?
    - What, if any, pretreatment is there for the systems? Describe...
    - If reverse osmosis is produced, what is the reject ratio...i.e., what percentage of water goes back to sewer and what percentage is used as process water?
    - Is water kept in a holding tank and drawn off as needed?
    - Or is it produced on an as-needed basis?
    - Is the water metered in? Out?

22. What solvents are used at this facility?
    - Where are they used? What unit operations?
    - What is the method of application? Dip tank, rag applied, or other?
    - Is there any discharge of solvents to sewer in any form? From the unit operation(s)? Final triple rinsing of containers prior to disposal? Describe...
    - If no solvents are discharged to sewer, explain what procedures are in place to prevent discharge.
    - Solvent Management Plan (SMP)?
    - When was SMP submitted?

23. What types of rinse tanks, rinsing procedures or configurations are in use at

this facility? Explain or describe...
- Running rinses
- Still rinses
- Countercurrent rinses
- Spray rinses
- Fog rinses
- Other

24. Is there adequate spill containment in the processing area(s)?
- If yes, describe.
- If no, describe.
- How is the sewer otherwise protected from leaks or spills in the process area(s) and throughout the facility? Describe in detail.

❑ COATINGS

Protective or decorative coatings or films are produced on metal surfaces by chromating, phosphating, immersion plating or coloring procedures. Sealing rinses are also used in conjunction with some coatings.

Chromating - A portion of the bare metal is converted to one of the components of the film by reaction with aqueous solutions containing hexavalent chromium and other active organic or inorganic compounds. Chromate conversion coatings are most frequently applied to the following metals:

- Zinc
- Cadmium
- Aluminum
- Copper
- Bronze
- Silver
- Magnesium
- Brass

The coatings can be applied by either electrochemical action or chemical immersion. These coatings are used for protective or decorative purposes or as a base for paint when the original material does not have good adhering properties for paint. Chromate conversion coatings are frequently applied to zinc or cadmium plated parts immediately following electrodeposition. Most chromate treatments employ proprietary solutions.

Passivating - Passivating is the treatment of metal with a strong oxidizing agent like nitric acid and/or sodium dichromates whereby an oxide coating is formed on the metal's surface. Passivation can be accomplished either by anodizing (see section on anodizing) or chromating (see above). Chromating, or chromate conversion coating, is also frequently referred to as alodizing or alodyning. Alodine is the trade-name for a proprietary coatings solution used to treat aluminum and the term is sometimes used synonymously with passivating and/or chromate conversion coating.

Phosphating - Phosphating is the treatment of iron, steel, zinc plated steel, and other metals by immersion in a dilute solution of phosphoric acid plus other reagents to produce an integral conversion coating on the surface.

The process is similar to chromating except that the new surfaces have non-metallic, non-conductive surfaces. Phosphating is done to prolong the life of paint finishes.

<u>Immersion Plating</u> - This is a chemical plating process in which a thin metal deposit is obtained by chemical displacement of the basis material. In immersion plating a metal will displace from solution any other metal that is below it in the electromotive series of elements. The less active metal will be deposited from solution while the more active metal (the item being plated) will be dissolved. This process is used to insure corrosion protection or as a preparation for painting or rubber bonding. It is mostly used for the following combinations:

- Tin on brass, copper, steel or aluminum
- Copper on steel
- Gold on copper or brass
- Nickel on steel

Preparation for immersion plating consists of an alkaline cleaning step and a pickling step. Pickling is the removal of scale, oxides, and other impurities from metal surfaces by immersion in an inorganic acid, usually sulfuric, hydrochloric, or phosphoric. Rate of removal varies inversely with concentration and temperature, with the usual concentration at 15% at approximately 100 C.

<u>Metal Coloring</u> - Metal coloring involves the application of color - particularly to copper, steel, zinc and cadmium - by a brief immersion of the piece to be colored in dilute aqueous solutions and are sometimes referred to as anodizing processes. The colored films produced are thin and delicate and often require a coat of lacquer. The solutions used vary widely. Examples are:

- Ammonium molybdate (black on zinc and aluminum)
- Black oxide process (with dichromate)
- Ammonium polysulfide (gray or silver)
- Copper carbonate and ammonia (black on brass)
- Black oxide process (black on steel, zinc, and cadmium)
- Ferric chloride/potassium ferricyanide (blue on aluminum)

- Potassium dichromate/nitric acid (Brown on cadmium)
- Potassium chlorate/nickel sulfate (Brown on copper)

<u>Sealing Operations</u> - The sealing step is included when the treated surface is porous and highly vulnerable to degradation. Sealing with a distilled or de-ionized water at about 100° C is the simplest form and also the preferred option for sealing dyed anodized aluminum. Where extreme corrosion resistance is required, a dichromate seal is commonly used. The other most common sealing process is the nickel/cobalt acetate sealing. This procedure provides more corrosion resistance than plain water sealing, but it is not as effective as the dichromate seal and can therefore provide a sealing option part way between the other two methods. An added of the nickel/cobalt acetate seal is the ability to use tap water in its make-up rather than the more expensive distilled or de-ionized water required with the plain water seal.

INSPECTOR'S QUESTIONS—COATINGS

1. When did coating operations begin at this facility? Day/month/year?

2. Is this facility an independent job shop, or are the materials being processed owned by the company? Describe if startup occurred prior to August 31, 1982.

3. What type of coating process is being performed at this facility? (Common types are chromating [chromate conversion], phosphating, immersion plating [displacement reaction] and metal coloring.)

4. For what purpose is the metal being coated? (May be post-plating for decoration or corrosion resistance, pre-painting to provide good adhesion properties for the paint, etc.)

5. Are there any preparatory steps prior to the coating step which involve a chemical treatment associated with rinse water? (Immersion plating commonly has an alkaline cleaning step and a pickling step prior to the plating operation.)

6. If an acid pickle step is involved, describe it... (What type of acid used?)

7. Are there any other metal bearing solutions present in the process tanks or from other chemical solutions used or other operations at the facility?

8. Are there any cyanide bearing solutions or waste streams present?
    - If yes, from which process(es)?
    - Where does it discharge? Describe...

9. Is any vapor degreasing done?
    - If yes, is the unit water cooled (one-pass), on a cooling tower or refrigerant type?
    - If one-pass, where does it discharge to sewer?
    - What type of solvent is used in the vapor degreaser?
    - How is the spent or contaminated solvent handled? Disposed of?

Describe...

10. What are the individual volumes and locations of the process tanks?
Describe...

11. Are alkaline cleaning or acid cleaning tanks present?
    - If yes, how frequently are these tanks batch discharged to the sewer?
    - What pretreatment occurs prior to discharge?
    - What are the individual volumes of all process tanks in facility - plating, acid, etc.?
    - How is bottom sludge handled?

12. How are spent or contaminated processing tanks handled?
    - If hauled, who hauls it? (Can check manifests.)

13. Are any tanks heated?
    - If yes, which ones?

14. Any rectifiers, compressors or similar equipment in use?
    - If yes, do any of them require cooling water?
    - Any one-pass cooling water?
    - If yes, volume (gallons per day)?
    - Where is the discharge point? Describe...
    - If water is not used for cooling equipment, describe the cooling practice.

15. Is there any routine washdown of the work area(s)?
    - Where is the discharge point? Describe...

16. Are there any floor drains in the process area?
    - Does their flow pass through a common interceptor point?
    - If no, where does it discharge? Describe...

17. Do all process streams flow to a common interceptor point?
    - If yes, where is that point?
    - If no, where are the various points?

18. Is there any pretreatment of waste water?
    - If yes, describe the various aspects in place and where located.

19. How is the routine disposal of hazardous wastes handled?
    - Who is the waste hauler?
    - What is the average volume of material disposed to landfill, recycled, etc.?
    - Where are the records kept? (Can ask to see manifests.)

20. Is there any water reuse in the facility?
    - If yes, where and what is it?

21. What type of rinse tanks, rinsing procedures, or configurations are in use at this facility? Explain or describe.
    - Running rinses
    - Still rinses
    - Countercurrent rinses
    - Spray rinses
    - Fog rinses
    - Other

22. Is there any reverse osmosis or de-ionized water production at this facility?
    - If yes, is it on a service? Who maintains the equipment?
    - What, if any, pretreatment is there for the system? Describe...
    - Is water kept in a holding tank and drained off as needed?
    - Or is it produced on an as-needed basis?
    - Is the water metered in? Out?
    - What is the reject ratio for the reverse osmosis system? I.e., what percentage of water goes back to the sewer as reject and what percentage is used as process water?
    - Where is the sample point for the wastewater discharge?
    - Does regulated wastewater from the coatings operations also flow to this location?

23. What solvents are used at this facility?
    - Where are they used? What unit operations?
    - What is the method of application? Dip tank, rag applied, or other?
    - Is there any discharge of solvents to sewer in any form? From the unit operation(s)? Final triple rinsing of containers prior to disposal? Explain...
    - If no solvents are discharged to sewer, explain what procedures are in

place to prevent discharge. Solvent Management Plan (SMP)?
- When was SMP submitted? Explain...

24. Is there adequate spill containment in the processing area(s)?
    - If yes, describe.
    - If no, describe.
    - How is the sewer otherwise protected from leaks or spills in the process area(s) and throughout the facility? Describe in detail.

❏ CHEMICAL MILLING AND ETCHING

Chemical milling is the process of shaping, machining, fabricating or blanking metal parts to specific design configurations and tolerances by controlled dissolution with chemical reagents or etchants. Highly concentrated solutions of sodium hydroxide are used in chemical milling.

Chemical etching is the process of removing relatively small amounts of metal to improve the surface condition of the basis metal or to produce a pattern such as is the case with printed circuit board manufacturing. Areas where no metal removal is desired are masked off by dipping, spraying or roll or flow-coating. Mask patterns can also be applied by the use of photosensitive resists, (principally used for etching printed circuit boards). After the masking step, the part is given an acid dip to activate the surface for etching. Etching/milling solutions include:

- Sodium Hydroxide - chemical milling
- Ferric chloride - chemical etching
- Nitric acid - chemical etching
- Chromic acid - chemical etching
- Sodium persulfate - chemical etching
- Ammonium persulfate - chemical etching
- Cupric chloride - chemical etching

Wastewaters typically contain the dilute etching solutions plus low to moderate concentrations of the particular metal being etched away.

INSPECTOR'S QUESTIONS—CHEMICAL MILLING/CHEMICAL ETCHING

1. When did milling or etching operations begin at this facility? Day/month/year?

2. Is this facility an independent job shop? Or are the materials being processed owned by the company? Describe...

3. What type of metal is being milled or etched? Describe...

4. What is the destination or use of the final etched or milled product?

5. What types of etching/milling chemicals are in use at this facility? Describe...

6. What individual operations are present at this facility as parts of the entire milling/etching process? (Cleaning, post-milling/etching steps and pickle steps are usually present)

7. What are the possible metal contaminants entering the waste stream, both from chemicals used in process tanks and/or from the metal parts themselves (etched away)? (Common etchants contribute copper, chromium and Iron to the waste stream. Metals being etched can be alloys containing any or all of the common metals excluding lead.)

8. What other metal bearing solutions are present in other process tanks or in other chemical solutions or from other operations at this facility?

9. Are there any cyanide bearing solutions or wastestreams present?
    - If yes, from which process(es)?
    - Where does it discharge? Describe...

10. Is there any metal coloring, coating, etc. associated with your etching/milling operation? Any organic dyes used? If yes, what type of discharge and where is the discharge point? What is the volume? Does it flow to a common point? If no, where does it go?

11. Is any vapor degreasing done prior to the etching/milling step?

- If yes, is the unit water cooled (one-pass), on a cooling tower or of the refrigerant type?
- If one-pass, where does it discharge to sewer?
- What type of solvent is used in the vapor degreaser?
- How is the spent or contaminated solvent handled? Disposed of? Describe...

12. What are the individual volumes and locations of the process and rinse tanks?

13. Are alkaline cleaning or acid cleaning tanks present?
    - If yes, how frequently are these tanks batch discharged to the sewer?
    - What pretreatment occurs prior to discharge?
    - What are the individual volumes of all process tanks in facility - plating, acid, etc.?
    - How is bottom sludge handled?

14. How are spent or contaminated processing tanks handled?
    - If hauled, who hauls it? (Can check manifests.)

15. Are any tanks heated?
    - If yes, which ones?

16. Any rectifiers, compressors or similar equipment in use?
    - If yes, do any of them require cooling water?
    - Any one-pass cooling water?
    - If yes, volume (gallons per day)?
    - Where is the discharge point? Describe...
    - If water is not used for cooling equipment, describe cooling practices.

17. Is there any routine washdown of the work area(s)?
    - Where is the discharge point? Describe...

18. Are there any floor drains in the process area?
    - Does their flow pass through a common interceptor point?
    - If no, where does it discharge? Describe...

19. Do all process streams flow to a common interceptor point?
    - If yes, where is that point?

- If no, where are the various points? Describe...
- Is there any co-mingling of single-pass or briny reject and categorically regulated wastewater at the sampling location?

20. Is there any pretreatment of waste water?
    - If yes, describe the various aspects in place and where located.

21. How is the disposal of hazardous wastes handled?
    - Who is the waste hauler?
    - What is the average volume of material disposed to landfill, recycled, etc.?
    - Where are the records kept? (Can ask to see manifests.)

22. Is there any water reuse in the facility?
    - If yes, where and what is it?

23. What type of rinse tanks, rinsing procedures, or configurations are in use at this facility? Explain or describe.
    - Running rinses
    - Still rinses
    - Countercurrent rinses
    - Spray rinses
    - Fog rinses
    - Other

24. Is there any reverse osmosis or de-ionized water production at this facility?
    - If yes, is it on a service? Who maintains the equipment?
    - What, if any, pretreatment is there for the system? Describe...
    - Is water kept in a holding tank and drained off as needed?
    - Or is it produced on an as-needed basis?
    - Is the water metered in? Out?
    - What is the reject ratio for the reverse osmosis system? I.e., what percentage of water goes back to the sewer as reject and what percentage is used as process water?
    - Where is the sample point for the wastewater discharge?
    - Does regulated wastewater from the chemical etching or milling also flow to this sampling location?

25. What solvents are used at this facility?
    - Where are they used? What unit operations?
    - What is the method of application? Dip tank, rag applied, or other?
    - Is there any discharge of solvents to sewer in any form? From the unit operation(s)? Final triple rinsing of containers prior to disposal? Explain...
    - If no solvents are discharged to sewer, explain what procedures are in place to prevent discharge. Solvent Management Plan (SMP)?
    - When was SMP submitted? Explain...

26. Is there adequate spill containment in the processing area(s)?
    - If yes, describe.
    - If no, describe.
    - How is the sewer otherwise protected from leaks or spills in the process area(s) and throughout the facility? Describe in detail.

❏ PRINTED CIRCUIT BOARDS

Printed Circuit Boards (PCB's or PC Boards) are made from nonconductive board materials such as plastic or glass. A circuit is created by sandwiching conductive metal, usually copper, on or between multiple layers of board in order to provide conductive circuits across and through the board. The board not only provides a surface for the application of a wiring path but also gives support and protection to the components it connects. Printed circuit boards find widespread use in such applications as business machines and computers, plus communication, microwave and home entertainment equipment. The industry employs three main production methods:

- Subtractive Process

- Additive Process

- Semi-Additive Process

• The Subtractive Production Method

The subtractive process derives its name from the large amount of material that is removed in order to make the circuit. The conventional subtractive process begins with a laminate board composed of a nonconductive material such as glass epoxy or phenolic paper. This board is first covered with a metallic foil, usually copper, and holes are drilled through it. After cleaning and surface preparation, the panel is plated entirely with electroless copper in order to deposit a uniform conductive layer over the entire board, including the inside surfaces of the holes.

At this point, the board can be plated in one of two ways.

1. Panel plating:

    - Entire board is electroplated with copper
    - A plating resist is applied and only the desired circuit is left exposed
    - Entire board is then immersion plated in a tin-lead bath to solder coat the circuits

2. Pattern plating:

    - A plating resist is applied and only the desired circuit is copper plated
    - Entire board is immersion plated in a tin-lead bath to solder coat the circuits

Following either process, the board now has the following layers.

1. Original foil covering
2. Electroless copper plate
3. Copper electroplate
    - on the board and circuit if _panel_ plated
    - on the circuit only if _pattern_ plated
4. Tin-lead solder on the circuit

Following the application of the solder plate by either method the plating resist is stripped off, exposing the copper in areas where the circuit is not required. This copper is then etched off. The circuit itself is protected by its coating of solder. After this etching step the solder is stripped from the tabs or fingers at the edges of the board. These areas are replated with other materials (usually gold, or nickel and gold) to meet customer specifications. This is referred to as <u>nickel tab plating</u>. When the nickel tab plating is complete the solder is reflowed to the circuit to seal and protect the underlying copper. Finally, boards are blanked and cut to size. See Figure L-3.

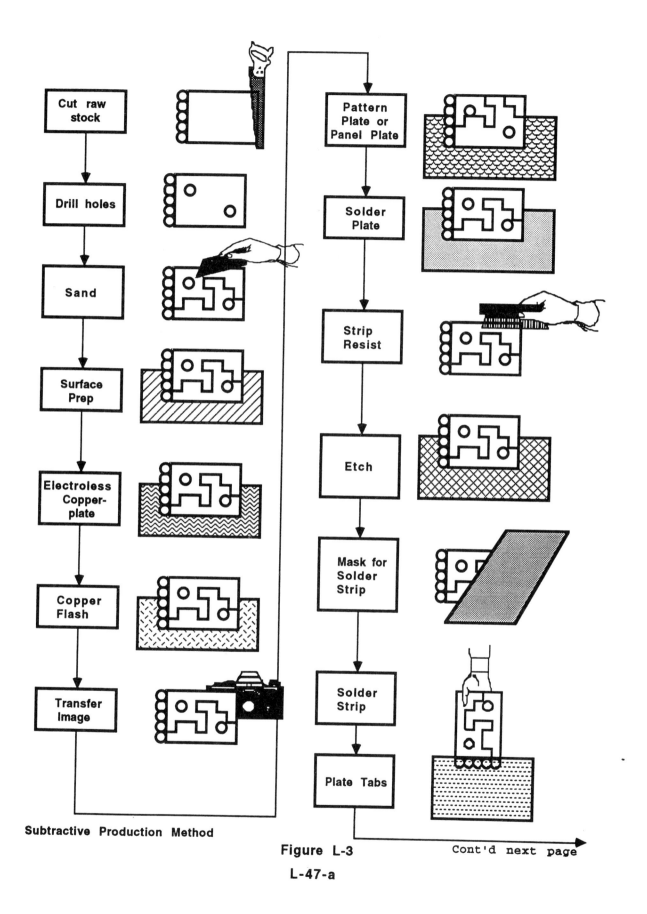

Figure L-3
Subtractive Production Method
L-47-a

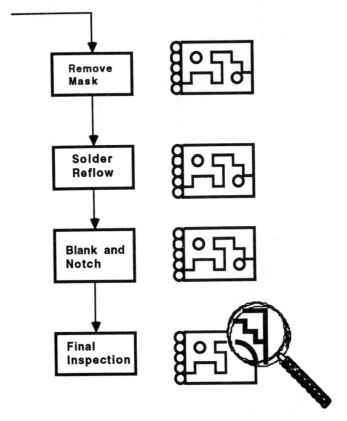

Subtractive Production Method

Figure L-3

- **The Additive Production Method**

    The additive process involves the deposition of plating material on the board in the pattern dictated by the circuit, rather than by removing metal already deposited to "expose" the circuit pattern (as in the subtractive process).

    The process begins with a bare board which may or may not be impregnated with a catalyst. Holes are then formed by drilling or punching. The next step is an adhesion promotion operation where an adhesive is applied and the surface is roughened or etched in order to make it microporous. The roughing or etching is required because of the large area that must be electroless plated. Usually a catalyst is applied after this roughening operation. Following this, the plating resist outlining the required circuit pattern is applied to the board in the non-circuit areas. An accelerator step, necessary for electroless plating, is then carried out and the board is immersed in the electroless copper bath. Following the copper deposition, the tabs are plated in the same manner as in the subtractive process. At this point different finishing steps may be applied such as an application of a protective coating to the board, etc.. See Figure L-4.

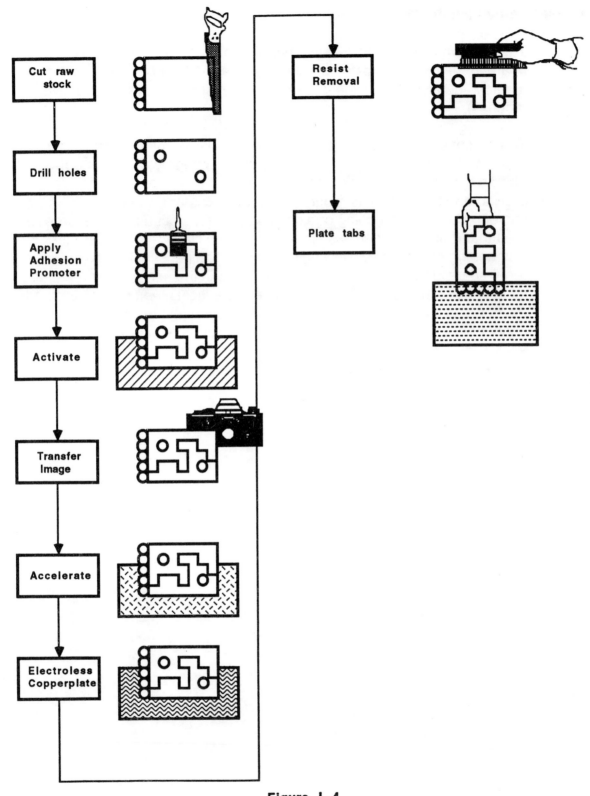

Figure L-4

- **The Semi-Additive Production Method**

    The semi-additive production process is a compromise between additive and subtractive processes. It is rarely found in industry. The primary difference between semi-additive and subtractive processes is that the semi-additive process begins with a bare board. This simply means that when etching occurs later on there is less copper to be removed. Adhesion promoters and surface preparation steps are extensively used as in the additive process, but the process is really a variation on the subtractive.

    Boards are produced in a wide variety of layer configurations and also with application differences, each of which requires process modifications. These are best identified under field conditions. For example, some shops panel plate in the following manner:

    - Drilled panels are electroless copper plated
    - Entire panel is copper electroplated to desired thickness
    - A negative image is applied with a photo etch resist to protect 'wanted straps or traces'
    - All undesired copper is etched off
    - All etch resist is stripped off
    - Molten solder is applied to the exposed straps and holes and leveled with hot air or liquid
    - Optional addition of a final step of flux and a liquid fuser

- **Process Descriptions/Waste Sources**

    All circuit board production, regardless of the type process used, is based on the following operations.

    - Cleaning and Surface Preparations (including photo/photo masking and resist operations)

    - Catalyst Application

    - Electroless Copper Plate

    - Electroplating

- **Cleaning and Surface Preparations**

    This is a crucial step in the printed circuit board production. For a board to be plated correctly without flaws, it must be properly cleaned and properly treated chemically.

    Following are the critical steps and the associated waste stream products:

    - Scrubbing......................................Copper particles

    - Alkaline cleaning............................Alkaline cleaning solutions with possibility of metal bearing sludges being produced

    - Etching/surface preparation........Sulfates
       Copper wastes, both liquids and sludges

    - Etching/Surface preparation.......Chromium (if additive/semi-additive process used)
       Metal bearing waste streams & sludges

    - Acid cleaning..................................Acid cleaning solutions with metals in solution and associated dilute acidic wastestreams from running rinses, dead rinses and similar operations

- **Catalyst Application**

    In the subtractive process, electroless copper clings readily to the copper clad board. For the additive and semi-additive processes, however, a catalyst must be used so that the copper will plate onto the bare board.

    Use of the catalyst involves two steps: application and acceleration. The application step involves interactions between palladium and tin. Acceleration extends the process and involves a second tin reaction.

- **Electroless Copper Plate**

    After the boards have been catalyzed they go into the electroless copper solution and are panel plated in the subtractive and semi-additive processes or pattern plated in the additive process.

    The electroless copper bath contains:

    - Copper salts (copper sulfate is most prevalent)
    - Formaldehyde
    - Chelating agents (usually tartrate or an EDTA compound)
    - Sodium hydroxide
    - Various polymers and amines

    Of particular note are the chelating agents. These agents serve to keep metal ions in solution and often make waste treatment difficult. See Table L-6, page L-24.

- **Electroplating**

    Electroplating is performed at several junctures in the production of printed circuit boards. It is employed in the actual buildup of the circuit (in the subtractive and semi-additive processes); it applies the tin-lead solder (etch resist and anti-corrosion layer) to the circuit; and it covers the tabs or fingers of all boards (gold, nickel).

    In order to build up the desired circuit in the subtractive (and semi-additive) processes, copper electroplating is used, followed by solder electroplating. The copper bath itself is usually one of four types:

    - cyanide copper
    - fluoborate copper
    - pyrophosphate copper
    - sulfate copper

    After the application of the copper electroplate, solder electroplate is applied. This serves a dual purpose. First, it acts as a mask during the etching process and second, protects the copper circuit from corrosion after final fabrication.

    Although it is not a type of electroplating but rather a metal finishing process, mention is made here of tin immersion plating. This is a displacement type of plating in which a tin solution with a chelating agent is employed. The tin displaces the copper which then goes into solution. The chelating agent is used to tie up the copper going into solution; the tin complexes only weakly. This is a process almost universally found in printed circuit board shops and is used mainly for rework.

- **Primary Concerns regarding PCB Manufacturers**

    - Are water allocation to various wet operations accurate and reasonable?

    - Are constituents in waste stream(s) identified?

        -Chemical makeup of process tanks
        -How tabs are plated
        -Any metals complexed with cyanide
        -What solvents are used...are any discharged?

    - Is one-pass cooling water or reverse osmosis brine reject co-mingling with a categorical waste stream(s)?

    - Have 'batch' discharges from unit operations been identified; i.e., alkaline soak tanks, acid tanks, still rinse tanks, etc.?

    - Have all process streams been identified and the flows traced to <u>individual</u> or a <u>common</u> sample point?

- **Wastewater**

    Wastewater is produced in the manufacturing of printed circuit boards from the following processes:

    - Silkscreening operations - The rinses following cleanup of screens to remove screening inks, solder masks, and screen cleaners

    - Surface preparation - The rinses following board scrubbing, alkaline cleaning, acid cleaning, etching, catalyst application and activation

    - Electroless plating - Rinses following the electroless plating step

    - Panel or Pattern plating - Rinses following acid cleaning, alkaline cleaning, copper plating and solder plating

    - Etching - Rinses following etching and solder brightening

    - Tab plating - Rinses following solder stripping, scrubbing, acid cleaning and nickel, gold or other plating operations

    - Immersion plating - Rinses following acid cleaning and immersion tin plating

    Additionally, water may be used for subsidiary purposes such as rinsing away spills, air scrubbing operations, equipment washing and accompanying the dumping of spent process solutions (batch discharge).

<u>Principal constituents potentially found in waste stream from the PC Board industry</u>:

- Acidity
- Copper
- Nickel

- Lead
- Gold
- Cyanide
- Toxic Organics (Ex: MEK, other solvents)
- Fluorides
- Chelating Agents
- Phosphorus

INSPECTOR'S QUESTIONS

INSPECTOR'S QUESTIONS—PRINTED CIRCUIT BOARD MANUFACTURING

1. When did Printed Circuit Board manufacturing operations begin at this facility? Day/Month/Year...

2. Is this facility an independent job shop, printed circuit board manufacturer, or are the materials being produced owned by the company? Describe...

3. What type of Printed Circuit Board is being manufactured? (Multilayer, etc.)
    - What type of facility is it?

4. What is the purpose of the board? (Include brief description.)

5. What individual operations are present at this facility (i.e., silkscreening, drilling, etching, etc.)?
    - List them.

6. What metals are being plated?
    - From what processes? Describe...

7. What other metal bearing solutions are present in process tanks or from other chemical solutions or operations at this facility? Describe...

8. Are any metals complexed with cyanide? If yes, which ones?

9. Is silkscreening and/or photo developing involved in your printed circuit board operation?
    - Where is/are that/those operation(s)?
    - What is the type and volume of discharge? Describe chemicals used...
    - Where is the discharge point?
    - If no discharge, describe how chemicals are handled
    - If yes, does it flow to a common point?
    - If no, describe...

10. Is any vapor degreasing done?

- If yes, is the unit water cooled (one-pass), on a cooling tower or refrigerant type?
- If one-pass, where does it discharge to sewer?
- What type of solvent is used in the vapor degreaser?
- How is the spent or contaminated solvent handled? Disposed of? Describe...

11. What are the individual volumes and locations of the process tanks? Describe...

12. Are alkaline cleaning or acid cleaning tanks present?
    - If yes, how frequently are these tanks batch discharged to the sewer?
    - What pretreatment occurs prior to discharge?
    - What are the individual volumes of all process tanks in facility - plating, acid, etc.?
    - How is bottom sludge handled?

13. How are spent or contaminated processing tanks handled?
    - If hauled, who hauls it? (Can check manifests.)

14. Are any tanks heated?
    - If yes, which ones?

15. Any rectifiers, compressors or similar equipment in use?
    - If yes, do any of them require cooling water?
    - Any one-pass cooling water?
    - If yes, volume (gallons per day)?
    - Where is the discharge point?
    - If water is not used for cooling equipment, describe cooling practices

16. Is there any routine washdown of the work area(s)?
    - Where is the discharge point? Describe...

17. Are there any floor drains in the process area?
    - Does their flow pass through a common interceptor point?
    - If no, where does it discharge? Describe...

18. Do all process streams flow to a common interceptor point?
    - If yes, where is that point?

- If no, where are the various points? Describe...
- Is one-pass or brine reject water co-mingling with categorically regulated wastewater at the sample point(s)?
- (See Combined Waste Stream Formula)

19. Is there any pretreatment of waste water?
    - If yes, describe the various aspects in place and where located.

20. How is the routine disposal of hazardous wastes handled?
    - Who is the waste hauler?
    - What is the average volume of material disposed to landfill, recycled, etc.?
    - Where are the records kept? (Can ask to see manifests.)

21. Is there any water reuse in the facility?
    - If yes, where and what is it?

22. What solvents are used at this facility?
    - Where are they used? What unit operations?
    - What is the method of application? Dip tank, rag applied, or other?
    - Is there any discharge of solvents to sewer in any form? From the unit operation(s)? Final triple rinsing of containers prior to disposal? Explain...
    - If no solvents are discharged to sewer, explain what procedures are in place to prevent discharge.

23. What type of rinse tanks, rinsing procedures, or configurations are in use at this facility? Explain or describe.
    - Running rinses
    - Still rinses
    - Countercurrent rinses
    - Spray rinses
    - Fog rinses
    - Other

24. Is there any reverse osmosis or de-ionized water production at this facility?
    - If yes, is it on a service? Who maintains the equipment?
    - What, if any, pretreatment is there for the system? Describe...
    - Is water kept in a holding tank and drained off as needed?

- Or is it produced on an as-needed basis?
- Is the water metered in? Out?
- What is the reject ratio for the reverse osmosis system? I.e., what percentage of water goes back to the sewer as reject and what percentage is used as process water?
- Where is the sample point for the wastewater discharge?
- Does regulated wastewater from the printed circuit board manufacturing operation also flow to this location?

25. Is there adequate spill containment in the processing area(s)?
    - If yes, describe.
    - If no, describe.
    - How is the sewer otherwise protected from leaks or spills in the process area(s) and throughout the facility? Describe in detail.

❏ ADDITIONAL 40 METAL FINISHING PROCESS OPERATIONS

Of the 46 process operations mentioned earlier, the six which are also regulated under 40 CFR Part 413 have been outlined in the electroplating section. The additional 40 process operations, regulated under 40 CFR Part 433, are listed below with a brief description of each process operation. For a listing of potential wastewater pollutants generated by metal finishing unit operations, see Table L-8, page L-60-a.

If any of the operations below occurs in conjunction with at least one of the first six electroplating operations it is subject to the Metal Finishing regulations. If a facility does not perform at least one of these six electroplating operations, with one of the process operations described below, the auxiliary process is excluded from Metal Finishing Regulations 40 CFR Part 433 and is subject to local limits only.

CLEANING

This operation involves the removal of oil, grease and dirt from the basis material using water with or without detergents or other dispersing agents. Acid cleaning is a process in which an acid is used with a wetting agent or detergent to remove oil, grease, dirt or oxide from the metal surface. Wastewater is produced from the 'still' or 'running' rinses post cleaning.

MACHINING

This operation involves the general process of removing stock from a workpiece by forcing a cutting tool through the workpiece, removing a chip of basis material. Machining operations incorporate the use of natural and synthetic oils for cooling and lubrication. Wastewater is produced from area washdown, spills, or batch dumping of spent cooling or lubricating oils.

GRINDING

This operation involves the process of removing stock from a workpiece by the use of a tool made of abrasive grains held together by a rigid or semi-rigid binder. Natural and synthetic oils are used for cooling and lubrication in many grinding operations. Wastewater is produced as a result of area washdown, spills, or batch dumping of spent cooling or lubricating oils.

POLISHING

This abrading operation is used to remove or smooth out surface defects (scratches, pits, tool marks, etc.) that adversely affect the appearance or function of a part. Area washdown or cleaning of equipment can produce regulated wastewater.

BARREL FINISHING (OR TUMBLING)

This operation is a controlled method of processing parts to remove burrs, scale, flash and oxides as well as to improve surface finish. Barrel finishing produces a uniformity of surface finish not possible by hand finishing and is generally the most economical method of cleaning and surface conditioning. Wastewater is generated by rinsing of parts following the finishing operation and by periodic dumping of process solutions.

BURNISHING

This operation involves the process of finish sizing or smooth finishing a workpiece (previously machined or ground) by displacement, rather than removal, of minute surface irregularities. Wastes may come from spills, leaks, process solution dumps and post-finish rinsing and could contribute common metals, precious metals and oily waste depending upon the basis material finished. In addition, sodium cyanide (NaCN) may be used as a wetting agent and rust inhibitor (for steel), thus also contributing cyanide to the waste stream.

IMPACT DEFORMATION

This operation involves the process of applying an impact force to a workpiece such that the workpiece is permanently deformed or shaped. Area washdown or cleaning of equipment could produce regulated wastewater.

PRESSURE DEFORMATION

This operation involves the process of applying force (at a slower rate than at impact force) to permanently deform or shape a workpiece. Area washdown or equipment cleaning can produce regulated wastewater.

SHEARING

This operation involves the process of severing or cutting a workpiece by forcing a sharp edge or opposed sharp edges into the workpiece stressing the material to the point of shear failure and separation. Area washdown or equipment cleaning can produce regulated wastewater.

HEAT TREATING

This operation involves the modification of the physical properties of a workpiece through the application of controlled heating and cooling cycles. Wastewater is generated through rinses, bath discharges, spills and leaks, and often contains the solution constituents as well as various scales, oxides and oils.

THERMAL CUTTING

This operation involves the process of cutting, slotting or piercing a workpiece using an oxyacetylene oxygen lance or electric arc cutting tool. Water may be used for rinsing or cooling of parts and equipment following this operation.

WELDING

This operation involves the process of joining two or more pieces of material by applying heat, pressure or both, with or without filler material, to produce a localized union through fusion or recrystallization across the interface. This operation is followed by quenching, cooling or annealing in a solution of water or emulsified oils. Wastewater is produced as a result of area washdown and batch dumping of quenching, cooling, or annealing tanks.

BRAZING

This operation involves the process of joining metals by flowing a thin, capillary thickness layer of nonferrous filler metal into the space between them.

Bonding results from the intimate contact produced by the dissolution of a small amount of base metal in the molten filler metal, without fusion of the base metal. The term brazing is used where the temperature exceeds 425º C (800º F). This operation is followed by quenching, cooling or annealing in a solution of water or emulsified oils. Wastewater is produced as a result of area washdown and batch dumping of quenching, cooling, or annealing tanks.

SOLDERING

This operation involves the process of joining metals by flowing a thin (capillary thickness) layer of nonferrous filler metal into the space between them. Bonding results from the intimate contact produced by the dissolution of a small amount of base metal in the molten filler metal, without fusion of the base metal. The term soldering is used where the temperature range falls below 425º C (800º F). This operation is followed by quenching, cooling or annealing in a solution of water or emulsified oils. Wastewater is produced as a result of batch dumping of the quenching, cooling, or annealing tanks, the use of post-solder running rinse, and from area washdown.

FLAME SPRAYING

This operation involves the process of applying a metallic coating to a workpiece using finely powdered fragments of wire, together with suitable fluxes, which are projected through a cone of flame onto the workpiece. This operation is followed by quenching, cooling or annealing in a solution of water or emulsified oils. Wastewater is produced as a result of batch dumping of the quenching, cooling, or annealing tanks and from area washdown.

SAND BLASTING

This operation involves the process of removing stock, including surface films, from a workpiece by the use of abrasive grains pneumatically impinged against the workpiece. Wastewater is produced as a result of area washdown or batch dumping of rinse tanks or the use of a running rinse step after the sand blasting step.

ABRASIVE JET MACHINING

This operation is a mechanical process for cutting hard brittle materials. It is similar to sand blasting but uses much finer abrasives carried at high velocities (500-3000 fps) by a liquid or gas stream. Wastewater can be produced through solution dumps, spills, leaks or washdowns of work areas and contributes to the common metals and oily waste types.

ELECTRICAL DISCHARGE MACHINING

This operation is a process which can remove metal from any metal with good dimensional control. The machining action is caused by the formation of an electrical spark between an electrode, shaped to the required contour, and the workpiece. Rinsing of machined parts and work area cleanups can generate wastewaters which also contain base materials.

ELECTROCHEMICAL MACHINING

This operation is a process based on the same principles used in electroplating except the workpiece is the anode and the tool is the cathode. Electrolyte is pumped between the electrodes and a potential applied which results in removal of the metal. In addition to standard chemical formulations, inorganic and organic solvents are sometimes used as electrolytes for electrochemical machining and with the basis material being machined, can enter waste streams via rinse discharges, bath dumps and floor spills.

ELECTRON BEAM MACHINING

This operation is a thermoelectric process whereby heat is generated by high velocity electrons impinging on part of the workpiece. At the point where the energy of the electrons is focused, it is transformed into sufficient thermal energy to vaporize the material locally and is generally carried out in a vacuum. Very limited impact on wastewater production except from area and equipment cleanup - usually considered a non-discharge area.

LASER BEAM MACHINING

This operation is the process whereby a highly focused monochromatic collimated beam of light is used to remove material at the point of impingement on a workpiece. Laser beam machining is a thermoelectric process with material removal largely accomplished by evaporation, although some material is removed in the liquid state at high velocity. Very limited impact on wastewater production except from area and equipment cleanup - usually considered a non-discharge area.

PLASMA ARC MACHINING

This operation is the process of material removal or shaping of a workpiece by a high velocity jet of high temperature ionized gas. A gas (e.g., nitrogen, argon or hydrogen) is passed through an electric arc causing it to become ionized and raised to temperatures in excess of 16,649° C (30,000° F). The relatively narrow plasma jet melts and displaces the workpiece material in its path. Very limited impact on wastewater production except from area and equipment cleanup - usually considered a non-discharge area.

ULTRASONIC MACHINING

This operation is a mechanical process designed to effectively machine hard, brittle materials. It removes material by the use of abrasive grains which are carried in a liquid between the tool and the work, and which bombard the work surface at high velocity. Very limited impact on wastewater production except from area and equipment cleanup - usually considered a non-discharge area.

SINTERING

This operation is the process of forming a mechanical part from a powdered metal by fusing the particles together under pressure and heat. The temperature is maintained below the melting point of the basis metal. Very limited impact on wastewater production except from area and equipment cleanup - usually considered a non-discharge area.

LAMINATING

This operation is the process of adhesive bonding layers of metal, plastic or wood to form a part. Water is not often used in this operation; however, occasional rinsing or cooling may occur in conjunction with laminating.

HOT DIP COATING

This operation is the process of coating a metallic workpiece with another metal to provide a protective film by immersion in a molten bath. Galvanizing (hot dip zinc) is the most common hot dip coating. Water is used for rinses following precleaning and sometimes for quenching after coating. Also area cleanup or batch dumping can produce substantial amounts of contaminated wastewater.

SPUTTERING

This operation is the process of covering a metallic or non-metallic workpiece with thin films of metal. The surface to be coated is bombarded with positive ions in a gas discharge tube, which is evacuated to a low pressure. Area washdown and equipment cleaning could produce wastewater - usually a non-discharge operation/area.

VAPOR PLATING

This operation is the process of decomposition of a metal or compound upon a heated surface by reduction or decomposition of a volatile compound at a temperature below the melting point of either the deposit of the basis material. Area washdown and equipment cleaning could produce wastewater - usually a non-discharge operation/area.

THERMAL INFUSION

This operation is the process of applying a fused zinc, cadmium or other metal coating to a ferrous workpiece by imbuing the surface of the workpiece with metal powder or dust in the presence of heat. Area washdown and equipment cleaning could produce wastewater - usually a non-discharge operation/area.

SALT BATH DESCALING

This operation is the process of removing surface oxides or scale from a workpiece by immersion of the workpiece in a molten salt bath or a hot salt solution. Molten salt baths are used to remove oxides from stainless steels and other corrosion-resistant alloys. These baths contain molten salts, caustic soda, sodium hydride and chemical additives. These contaminants (and a small amount of base material and oils) enter wastewater streams through rinsing, spills, leaks, batch dumps of process solutions and improper handling of sludge produced by the process. Area washdown can contribute contaminants to the wastestream as well.

SOLVENT DEGREASING

This operation is a process for removing oils and grease from the surface of a workpiece by the use of organic solvents such as aliphatic petroleums, aromatics, oxygenated hydrocarbons, halogenated hydrocarbons and combinations of these classes of solvents. These contaminants can enter wastewater streams and contribute to the toxic organic waste load. Usually low levels enter the waste stream because of post-degreasing rinse step. High levels are encountered when bad house keeping or inappropriate disposal methods are practiced.

PAINT STRIPPING

This operation is the process of removing an organic coating from a workpiece. The stripping of such coatings is usually performed with caustic, acid, solvent or molten salt. The stripping wastes can contain any of the constituents of the paint being removed, as well as a small amount of the basis material beneath the paint and the constituents of the stripping solution. Wastes are primarily generated by a rinsing step following the stripping step. Small amounts of emulsified oils can also be present in this waste stream but the major constituent of concern is usually an organic solvent such as methylene chloride.

PAINTING

This operation is the process of applying an organic coating to a workpiece. Wastewater is produced during cleanup of equipment, area washdown, or inappropriate dumping of unused or contaminated materials to the waste stream.

ELECTROSTATIC PAINTING

This operation involves the application of electrostatically charged paint particles to an oppositely charged workpiece followed by thermal fusing of the paint particles to form a cohesive paint film. Wastewater is produced during area washdown or as a result of equipment cleanup.

ELECTROPAINTING

This operation is the process of coating a workpiece by either making it anodic or cathodic in a bath that is generally an aqueous emulsion of the coating material. Electropainting is used primarily for primer coats because it gives a fairly thick, highly uniform, corrosion resistant coating in relatively little time. Ultrafiltration is used in connection with electropainting to concentrate paint solids. Wastewater is produced as a result of area and equipment cleaning.

VACUUM METALIZING

This operation is the process of coating a workpiece with metal by flash heating metal vapor in a high-vacuum chamber containing the workpiece. The vapor condenses on all exposed surfaces. Wastewater is produced as a result of area and equipment cleanup.

ASSEMBLY

This operation involves the fitting together of previously manufactured parts or components into a complete machine, unit of a machine, or structure. Limited production of wastewater except during area washdown.

CALIBRATION

This operation involves the application of thermal, electrical or mechanical energy to set or establish reference points for a component or complete assembly. Limited production of wastewater except for area washdown. Usually only oily wastewater produced.

TESTING

This operation involves the application of thermal, electrical or mechanical energy to determine the suitability or functionality of a component or complete assembly. Leak testing, final washing (automobiles, etc.) and test area washdowns enter wastestreams and may contain oils and fluids used at testing stations as well as heavy metal contamination derived from the component being tested.

MECHANICAL PLATING

This operation is the process of depositing metal coatings on a workpiece via the use of a tumbling barrel, metal powder and usually glass beads for the impaction media. The operation is subject to the same cleaning and rinsing operations that are applied before and after the electroplating operation. Wastewater produced as a result of batch dumping, processing tanks, running rinse steps and area washdown.

Table L-7

POTENTIAL WASTEWATER POLLUTANTS GENERATED BY METAL FINISHING PROCESS OPERATIONS

Unit	Operations	Metals	Hexavalent Chromium	Cyanide	Oils	Toxic Organics	Zero Discharge
1	Electroplating	X	X	X		X	
2	Electroless Plating	X	X			X	
3	Anodizing	X	X			X	
4	Conversion Coating	X	X	X		X	
5	Etching/Chemical Milling	X	X	X		X	
6	Printed Circuit Board Manufacture	X				X	
7	Cleaning	X	X	X	X	X	
8	Machining	X			X		
9	Grinding	X			X		
10	Polishing	X			X		
11	Barrel Finishing (Tumbling)	X	X		X	X	
12	Burnishing	X			X	X	
13	Impact Deformation	X				X	
14	Pressure Deformation	X				X	
15	Shearing	X				X	
16	Heat Treating	X			X	X	
17	Thermal Cutting	X					
18	Welding	X					
19	Brazing	X					
20	Soldering	X					
21	Flame Spraying	X					
22	Sand Blasting	X					
23	Other Abrasive Jet Machining	X					
24	Electric Discharge Machining	X			X		
25	Electrochemical Machining	X			X	X	
26	Electron Beam Machining						X
27	Laser Beam Machining						X
28	Plasma Arc Machining						X
29	Ultrasonic Machining						X
30	Sintering						X
31	Laminating	X					
32	Hot Dip Coating	X					
33	Sputtering						X
34	Vapor Plating						X
35	Thermal Infusion						X
36	Salt Bath Descaling	X			X		
37	Solvent Degreasing	X			X	X	
38	Paint Stripping	X			X	X	
39	Painting	X				X	
40	Electrostatic Painting	X	X			X	
41	Electropainting	X				X	
42	Vacuum Metalizing						X
43	Assembly	X			X	X	
44	Calibration						X
45	Testing				X		
46	Mechanical Plating	X	X				

❑ SOURCES OF WATER POLLUTION

DRAG-OUT: Contaminants in the discharge from electroplating shops and metal finishing industries originate in several ways. The most common source of pollution is from "drag-out," which is plating solution that clings to the work piece and contaminates the rinse water. The amount of pollutants contributed by drag-out is a function of many factors including the design of the racks or barrels carrying the parts to be plated and the shape of the parts themselves, as well as the holding time over the process tank prior to rinse.

RINSE WATER: Large volumes of rinse water are usually needed to clean the drag-out from the work. Rinsing actually serves two purposes:

1. It cleans the part

2. It protects subsequent process baths from "drag-in" contamination

Because of high flow rates used in conventional rinsing techniques, rinse waters are contaminated with relatively dilute concentrations of process solutions. Typically, rinse waters that follow plating solutions contain between 15 and 100 milligrams per liter (mg/l) of the metal being plated.

USED/SPENT PROCESS SOLUTIONS: Platers discard spent cleaners, acids and bright dips. Although these solutions are not usually made up of metals, it is not uncommon to find cyanide and heavy metals in concentrations of several thousand milligrams per liter in these solutions. This contamination is caused by drag-in from previous process cycles and from metals leached from the work by the process chemicals. Plating solutions and other process chemicals containing high metal concentrations are rarely discarded. Instead, they are decontaminated or rejuvenated in place so they are usually not a hazardous waste problem.

ACCIDENTAL SPILLS, LEAKS, AND DRIPS OF PROCESS SOLUTIONS: These also contribute to effluent contamination.

ADDITIONAL POLLUTION SOURCES: These include:

- Sludges from the bottoms of plating baths
- Backwash from plating tank filter systems
- Stripping solutions
- Batch discharges from static rinse tanks, spent process tanks (alkaline, acid or process operations)

NOTE: Many facilities operate several plating lines such as zinc, copper, nickel, cadmium, and chromium. The rinse waters discharged from each line are usually combined in a common pipe or floor trench, and the concentrations of the individual metals from each process are diluted in the entire volume of the shop's wastewater, usually to less than 50 mg/l each.

The better technique is to isolate each plating rinse stream so that individual waste streams can be either reclaimed, reused, or conventionally pretreated. This type of selectivity allows the plater to conduct his operations more economically in that less water is used to rinse the processed parts, less chemistry is used to process the parts, and ultimately less waste of all types is produced, thereby eliminating a large, ongoing expense.

❑ WASTEWATER TECHNOLOGIES FOR WATER CONSERVATION/ WASTE MINIMIZATION

During the inspection of the facility, some of the following variations in rinsing technique may be observed:

- Methods to Reduce Water Consumption

    Δ Restricted pipe orifice sprays or mists to reduce flow (fog rinses)

    Δ Manual operation of water supply (might actually increase water consumption, depending on how used)

    Δ Intermittent rinsing of work

    Δ Static rinse tanks

    Δ Countercurrent rinse tanks

    Δ Series rinse tanks

    Δ Air agitation in rinse tanks to improve rinsing efficiency

    Δ Chemical addition to improve rinsing efficiency

    Δ Automatic discharge controls on rinse tanks

    Δ Clean water addition through automatic measurement of conductivity (might actually increase water consumption depending on how used)

    Δ Drag-out system; recycles rinse water into plating baths, generally only for baths that have a high evaporation rate and need supplemental water

Δ Proper drainage of rinsed material back into rinse tank

  Δ Use of demineralized water to minimize salt build up in rinse tank

- Methods to Reduce Discharge of Contaminants at the Source

  Δ Treatment of process solutions to extend usefulness

  Δ Product substitution (Alteration of production process such as noncyanide cleaners)

  Δ Use of spray rinses or "air knife" system on material from plating tanks

  Δ Use of minimum chemical concentrations that will obtain quality products

  Δ Use of spent baths as treatment reagents

  Δ For an expanded explanation of the above, see <u>Treatment Technologies</u> section.

❏ Application of the Combined Waste Stream Formula

The Combined Waste Stream Formula (CWF) is used to calculate appropriate limits for wastewater in which process wastestreams are mixed with other regulated, unregulated or dilute waste streams, resulting in a mixed wastestream. The CWF is applied to the mixed wastestream to account for the presence of the contributing streams.

The following definitions are important to the proper use of the Combined Wastestream Formula.

Definitions For the Application of the Combined Waste Stream Formula

*Note: Definitions apply to individual pollutants. A wastestream from a process may be "regulated" for one pollutant and "unregulated" for another

<u>Regulated Process Wastestream</u> - an industrial process wastestream regulated by National Categorical Pretreatment Standards

<u>Unregulated Process Wastestream</u> - an industrial process wastestream that is not regulated by a categorical standard

<u>Dilute Wastestream</u> - boiler blowdown, cooling tower blowdown, noncontact cooling water, sanitary wastewater, and Paragraph 8 excluded wastestreams containing none of the regulated pollutant or only trace amounts of it

<u>Concentration-based Limit</u> - a limit based on the relative

strength of a pollutant in a wastestream, usually expressed in mg/l (lb/gal).

Mass-based Limit - a limitation based on the actual quantity of a pollutant in a wastestream, usually expressed in mg/some unit of production for a given operation such as square meter (lb/square foot per operation).

Combined Wastestream Formulas

Alternative Concentration Limit Formula

$$C_t = \frac{\sum_{i=1}^{N} C_i F_i}{\sum_{i=1}^{N} F_i} \times \frac{F_t - F_d}{F_t}$$

C_t - alternative concentration limit for the pollutant

C_i - Categorical Pretreatment Standard concentration limit for the pollutant in regulated stream i

F_i - average daily flow (at least 30 day average) of regulated stream i

F_d - average daily flow (at least 30 day average of dilute wastestream(s)

F_t - average daily flow (at least 30 day average) through the combined treatment facility (including regulated, unregulated and dilute wastestreams)

N - total number of regulated streams

Alternate Mass Limit Formula

$$M_t = \sum_{i=1}^{N} M_i \times \frac{F_t - F_d}{\sum_{i=1}^{N} F_i}$$

M_t - alternative mass limit for the pollutant

M_i - Categorical Pretreatment Standard mass limit for the pollutant in regulated stream i

F_i - average daily flow (at least 30 day average) of regulated stream i

F_d - average daily flow (at least 30 day average) of dilute wastestream(s)

F_t - average daily flow (at least 30 day average) through the combined treatment facility (including regulated, unregulated and dilute wastestreams)

N - total number of regulated streams

❏ COMPLIANCE DATES

The pollutants that are regulated by the Metal Finishing Standards and the compliance dates for existing sources for each of these pollutants are shown below:

POLLUTANT PARAMETER	COMPLIANCE DATE FOR EXISTING SOURCES
Metals [Cadmium (T), Chromium (T), Copper (T), Lead (T), Nickel (T), Silver (T), and Zinc (T)]	February 15, 1986
Cyanide (T)	February 15, 1986
Total Toxic Organics	June 30, 1984 (interim limit)
	February 15, 1986 (final limit)

Section M

TREATMENT TECHNOLOGIES

Introduction

The treatment technologies reviewed in this section do not represent all known treatment technologies but are instead the most commonly used physical-chemical treatment methods. Most of the technologies are designed to remove or destroy a single pollutant, e.g., cyanide, or a related family of pollutants, such as heavy metals or greases and oils. Therefore, if an industrial site has several pollutants in their wastestream, their system will have to combine several different aspects of treatment technology to satisfactorily treat their effluent. This is the usual condition at industries where any metal finishing of consequence is done. It is a very rare facility that can satisfactorily, and legally, meet their permitted effluent limits without some type of wastewater treatment prior to discharge into the sanitary sewer.

The following specific technologies are included:

- Cyanide Destruction Technologies
- Metals Removal Technologies - Traditional Type
- Solvent Recovery and Removal Technologies
- Activated Carbon Adsorption
- Solids Removal
- Electrolytic Metal Recovery
- Reverse Osmosis
- Ion Exchange
- Crystallization
- Electrodialysis

- ❏ pH Neutralization
- ❏ Oxidation/Ozonation
- ❏ Equalization
- ❏ Grease and Oil Control Systems/Devices
- ❏ Rinsing Techniques

❏ CYANIDE DESTRUCTION TECHNOLOGIES

Present cyanide treatment processes demonstrated to be effective are based upon two fundamental approaches, chemical oxidation and high pressure and temperature techniques. Chemical oxidation is a reaction in which one or more electrons are transferred from the chemical being oxidized to the chemical initiating the transfer (oxidizing agent); as a result of the valence change, the oxidized substance can then react to form a more desirable compound. The other treatment is the application of high temperature and pressure to break down chemical bonds resulting in more tolerable substances (e.g., CO_2 and NO_2).

- **Chemical Oxidation**

    Δ Chlorination: Destruction of cyanide by oxidation either with chlorine gas under alkaline conditions or with sodium hypochlorite is a very common method of treating industrial wastewaters. Although more costly, sodium hypochlorite is less hazardous and is simpler to handle. Oxidation is approximated by a two-step chemical reaction. This reaction requires 30 minutes, during which time the cyanide is oxidized to cyanate completely and rapidly at a pH of 9.5 to 10.0. This oxidation produces a reduction in volatility and a thousandfold reduction in toxicity.

    Since cyanate may revert to cyanide under some conditions, additional chlorine is added to oxidize cyanate to carbon dioxide/bicarbonate. The excess chlorine also serves to break down cyanogen chloride, a highly toxic intermediate compound formed during the oxidation of cyanate.

    Because of some of the advantages of the chlorination process, this technology has received widespread application in the chemical industry as a whole. First, it is a relatively low cost system and does not require complicated equipment. It also fits well into the flow scheme of a wastewater treatment facility. The process will operate effectively at ambient conditions and is well suited for automatic operation, minimizing labor requirements. This technique is used by pharmaceutical processes manufacturers who use cyanide in

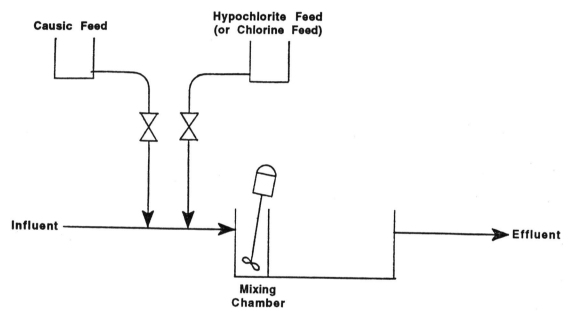

CYANIDE DESTRUCTION SYSTEM - CHLORINATION
Figure M-1

chemical synthesis.

Δ Ozonation: Ozone (allotropic form of oxygen is a good oxidizing agent and can be used to treat process wastewaters that contain cyanide. In fact, ozone oxidizes many cyanide complexes (for instance iron and nickel complexes) that are not broken down by chlorine. Ozonation is primarily used to oxidize cyanide to cyanate.

Oxidation of cyanide by ozone to cyanate occurs in about 15 minutes at a pH of 9.0 to 10.0. The reaction is almost instantaneous in the presence of traces of copper.

Oxidation of cyanate to the final end products, nitrogen and bicarbonate, is a much slower and more difficult process unless catalysts are present. Since ozonation will not easily effect further oxidation of cyanate, it is often coupled with such independent processes as dialysis or bio-oxidation.

The ozonation treatment process is beginning to receive more usage. Its initial applications in treating metal finishing wastewater have shown it to be quite effective for cyanide removal. Like chlorination, the ozonation process is well suited to automatic control and will operate effectively at ambient conditions. Also, the reaction product (oxygen) is beneficial to the treated wastewater.

- **High Pressure/Temperature Technique**

    - Alkaline Hydrolysis: In this process, a caustic solution is added to the cyanide-bearing wastewaters to raise the pH to between 9.0 and 12.0. Next, the wastewater is transferred to a continuous reactor where it is subjected to temperatures of about $165^{\circ}C$ to $185^{\circ}C$ and pressured from approximately 90 to 100 psi. The breakdown of cyanide in the reactor generally takes about 1.5 hours.

    The absence of specific chemical reactants in this process eliminates procurement, storage and handling problems. As with other cyanide processes, alkaline hydrolysis is well suited to automatic control.

❏ METALS REMOVAL TECHNOLOGIES - TRADITIONAL TYPES

Proven metals treatment technologies are based upon precipitation and filtration. The three primary treatment technologies for metals removal are chemical/chromium reduction, alkaline/hydroxide precipitation, and sulfide precipitation. A brief discussion of each of these technologies follows, as well as a short look at other methods which deserve mention.

- **Chemical/Chromium Reduction**: Chromium and some other metals must be reduced from their high valence states before they can be precipitated. This is accomplished by chemical reduction, a reaction in which one or more electrons are transferred from the chemical initiating the transfer (reducing agent) to the chemical being reduced.

  The primary application of chemical reduction in the treatment of industrial wastewater is the reduction of hexavalent chromium to trivalent chromium. Chromium is a common metal contaminant in the industry and its chemical reduction is employed as an in-plant treatment. The reduction enables the trivalent chromium in conjunction with other metal salts to be separated from solution by precipitation. (See Figure M-2.)

  Among the most common and useful of the chemical reduction agents are:

  - △ Sulfur Dioxide
  - △ Sodium Bisulfite
  - △ Sodium Metabisulfite
  - △ Ferrous Sulfate

  Chemical reduction has been used quite successfully to treat large concentrations of hexavalent chromium (e.g., from metal finishing operations). This method is well suited to automatic control and may be used when conditions are ambient. Careful pH control is required for effective reduction.

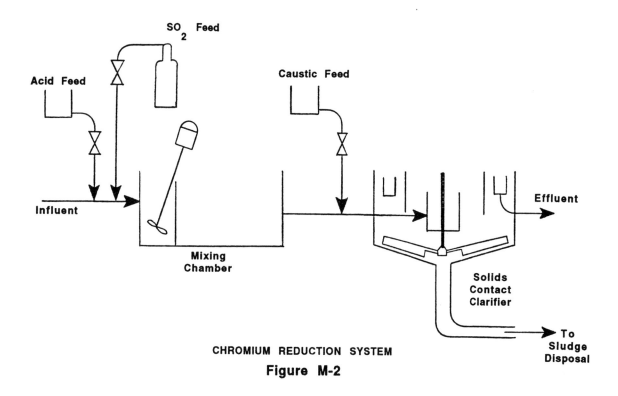

CHROMIUM REDUCTION SYSTEM
Figure M-2

- **Alkaline/Hydroxide Precipitation**: Alkaline precipitation is a classic technology being used by many industries. (See Figure M-3.)

The solubility of metal hydroxides, in most cases, is a function of pH and therefore the success of metal hydroxide precipitation treatment is heavily dependent on the pH level of the solution. In order to achieve optimum formation of solid metal hydroxides the pH of the wastewater must be adjusted to the range (usually moderately alkaline) found to be most effective for the metals involved. This is accomplished by measured addition of lime to the wastewater with concurrent pH monitoring.

Following the attainment of optimum pH conditions the solid metal hydroxides are coagulated (using coagulating agents) in a clarifier and deposited as sludge. Proper clarifier design and good coagulation are important prerequisites for efficient metals removal by alkaline precipitation.

There are several advantages to the use of alkaline precipitation. Above all, it is a well demonstrated wastewater treatment technology. It is well suited to automatic control and will operate at ambient conditions. Also, in many instances, preceding treatment steps adjust the waste (especially pH) to aid the alkaline precipitation process. The end result is that the costs associated with this technology may be substantially lower than those for other processes. However, this method is subject to interference when mixed wastes are treated. In addition, this process generates relatively high quantities of sludge that also require disposal. Generally, the use of caustic soda does not produce the large volume of sludge typical of treatment with lime. Also, because caustic soda is easily handled (as a drummed liquid) and does not produce much sludge, it has better applicability to small systems. The major disadvantage of caustic soda is its cost. Even though the treatment system installation may be less costly for a caustic soda system, the operation costs may more than offset this, especially in large systems.

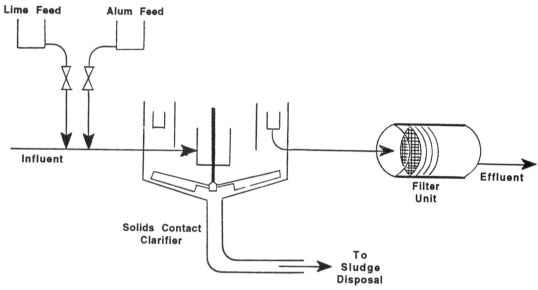

METALS REMOVAL SYSTEM - ALKALINE PRECIPITATION
Figure M-3

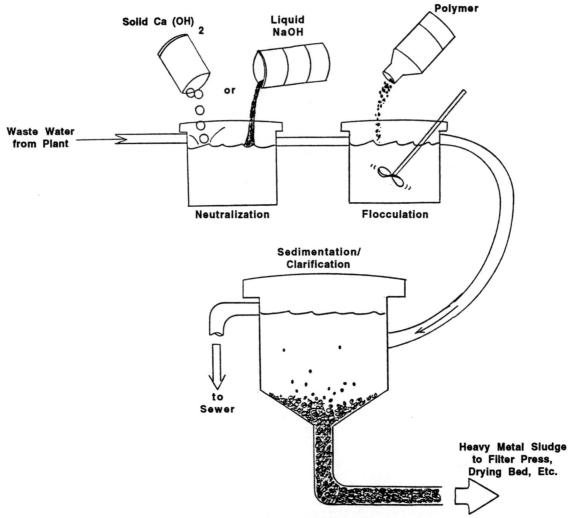

TRADITIONAL HYDROXIDE METAL PRECIPITATION SYSTEM
Not Recommended for Complexed Metals or Hexavalent Chromium

Figure M-4

CONVENTIONAL TREATMENT SCHEME

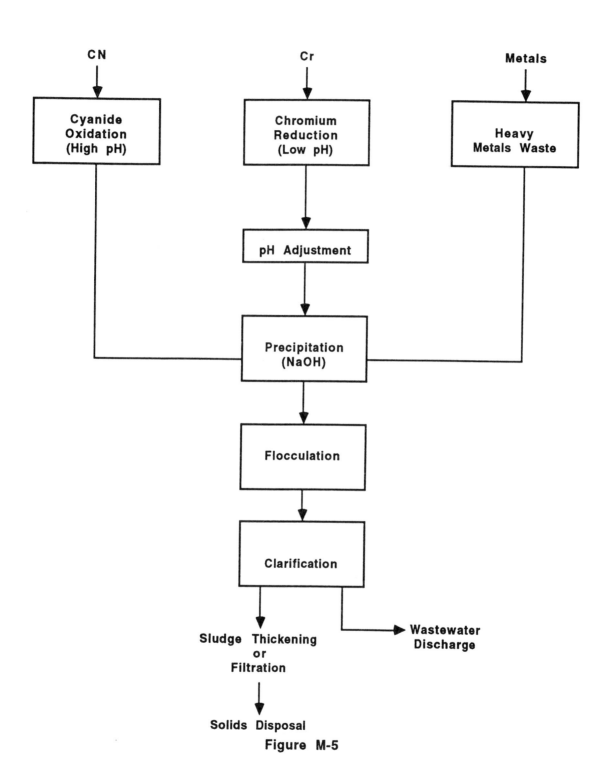

Figure M-5

- **Sulfide Precipitation**: In this process, heavy metals are removed as a sulfide precipitate. Sulfide is supplied by adding a vary slightly soluble metal sulfide that has a solubility somewhat greater than that of the sulfide of the metal to be removed. Two methods of sulfide precipitation are in common use. The first uses a soluble sulfide such as sodium sulfide, calcium polysulfide, or sodium hydrosulfide. In this method, sulfide solution is added to a waste solution with the pH maintained above eight. Careful management of this type of system is critical, and generally the sulfide is added by an automatic controller. A well operated sulfide precipitation system can remove metals to a very low level.

The second method uses a slightly soluble sulfide such as ferrous sulfide as a source of sulfide ions. The sulfide can be added as a powder or in a slurry. Most metal sulfides are less soluble than ferrous sulfide; consequently, the metal will precipitate as the iron dissolves. Because the reaction is performed at a pH greater than eight, the iron will subsequently precipitate as iron hydroxide.

In both methods, the minimum solubilities of the dissolved metals occur over a narrow pH range. Therefore, pH control is critical in sulfide precipitation processes.

A major advantage of sulfide precipitation is that hexavalent chromium is reduced to the trivalent state and then precipitated as chromium hydroxide without an additional reagent. This eliminates the two-step reduction required where the precipitation is accomplished with hydroxides.

The disadvantage of the soluble sulfide method is that in the presence of excess sulfide, some hydrogen sulfide gas can be produced.

The process is applicable for treatment of all heavy metals. The process equipment requirement includes a pH adjustment tank, a precipitator, a filter, and pumps to transport the wastewater. Pollutant levels after treatment with sulfide precipitation are very similar to the pollutant levels after alkaline precipitation.

❏ SOLVENT RECOVERY AND REMOVAL TECHNOLOGIES

Solvents are used extensively in the pharmaceutical manufacturing industry. Because such materials are expensive, most manufacturers try to recover them in order to purify them for reuse whenever possible. Solvent recovery operations typically employ such techniques as decantation, evaporation, distillation, and extraction. The feasibility and extent of recovery purification are governed largely by the quantities involved and by the complexity of solvent mixtures to be separated. If recovery is not economically practicable, the used solvents may have to be disposed of by means of incineration, landfilling, deep-well injection, or contract disposal.

Even when an effort is made to recover solvents, some wastewater contamination can be expected. Removal of small quantities of organic solvents from the segregated wastewater can be accomplished by such techniques as stream stripping or carbon adsorption. Further removal of solvents from combined end-of-pipe wastewater may result from biological treatment or from surface evaporation in the treatment process.

❑ ACTIVATED CARBON ADSORPTION

Adsorption is defined as the adhesion of dissolved molecules to the surface of solid bodies with which they are in contact. Granular activated carbon particles have two properties that make them effective and economical adsorbents:

√ A high surface area per unit volume

√ A high hardness value

The adsorption process typically is preceded by preliminary filtration or clarification to remove insolubles. Next, the wastewaters are placed in contact with carbon so adsorption can take place. Normally, two or more beds are used so that adsorption can continue while a depleted bed is reactivated. Reactivation is accomplished by heating the carbon between $870°$ C to $980°$ C to volatilize and oxidize the adsorbed contaminants. Oxygen in the furnace is normally controlled at less than one percent to avoid loss of carbon by combustion. Contaminants may be burned in an afterburner.

Carbon adsorption is particularly applicable in situations where organic material in low concentrations not amenable to treatment by other technologies must be removed from wastewater.

The equipment necessary for an activated carbon adsorption treatment system consists of a preliminary clarification and/or filtration unit to remove the bulk of the solids, two or three columns packed with activated carbon, and pumps and piping. When on-site regeneration is employed, a furnace, quench tanks, a spent carbon tank, and a reactivated carbon tank are generally required. Contract regeneration at a central location is a frequent commercial practice.

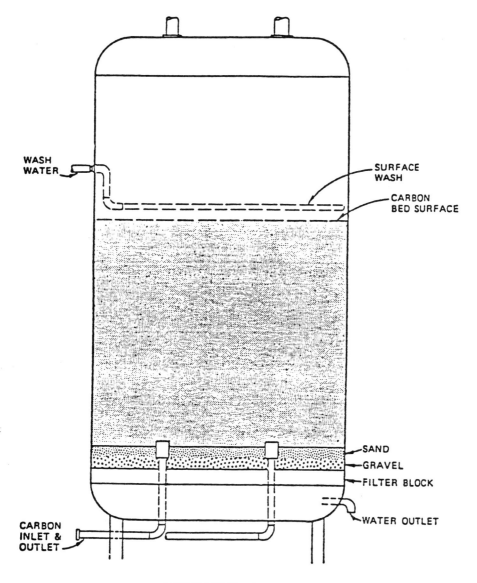

ACTIVATED CARBON ADSORPTION UNIT

Figure M-6

❑ SOLIDS REMOVAL

Removal of solids from wastewater can occur at several points in a treatment sequence. Grit removal by screening, filtration, or sedimentation is sometimes done as a preliminary step. The removal of sludge and other solids by means of clarification, filtration, or a special operation such as flotation can be done as a single solids removal step or in combination with the preliminary screening or filtration step.

- **Clarification**: Clarification is a method of removing suspended or colloidal solids by means of gravity sedimentation. Since the settling rate of suspended solids is dependent on particle size and density (the smaller the particle size and the closer the density to that of water, the slower the settling rate), flocculant or coagulant aids sometimes must be added to promote bridging between particles and to render them more settleable. A slow settling rate and the stability of colloidal mixtures make chemical destabilization and agglomeration of colloids/suspensions necessary.

Clarifiers are usually large containment vessels that have a continuous water throughput. A conventional clarification system utilizes a rapid mix tank to mix chemicals with the entering water; the wastewater is then subjected to slow agitation. Provision for the removal of settled solids is also a necessary part of the system.

It is not unusual to see clarifiers used without the chemical addition step. They perform some solids settling and also equalization when used in this way.

- **Filtration**: Another basic solids removal technology in water and wastewater treatment is filtration. The most common filtration system is the conventional gravity filter. It normally consists of a deep bed of granular media in an open-top tank. The direction of flow through the filter is downward and the flow rate is dependent solely on the hydrostatic pressure of the water above the filter bed. Another type of filter is the pressure filter. In this case, the basic approach is the same as a gravity filter, except the tank is enclosed and pressurized. Silica sand, anthracite coal, garnet, and similar granular inert materials are among the most common media used in this technology, with gravel serving as a support material. These media may be used separately or in various combinations. Multimedia filters may be arranged in relatively distinct layers by balancing the forces of gravity, flow, and buoyancy of the individual particles. This is accomplished by selecting appropriate filter flow rates, media grain size and media densities.

As wastewater is processed through the filter bed, the solids collect in the spaces between the filter particles. Periodically, the filter media must be cleaned. This is accomplished by backwashing the filter (reversing the flow through the filter bed). The flow rate for backwashing is adjusted in such a way that the bed is expanded by lifting the media particles a given amount. This expansion and subsequent motion provides a scouring action which effectively dislodges the entrapped solids from the media grain surfaces. The backwash water fills the tank up to the level of a trough below the top lip of the tank wall. The backwash is collected in the trough, fed to a storage tank, and recycled into the waste treatment stream. The backwash flow is continued until the filter is clean.

- **Flotation**: Flotation is an optional method of clarification utilized to treat some industrial waste in which the suspended solids have densities less than that of water. Air-assisted flotation may be applied to some systems with solids slightly heavier than water. As with conventional clarifiers, flocculants are frequently employed to enhance the efficiency of flotation.

Dissolved air flotation (DAF) is frequently used with certain oily wastes following the addition of a flocculating agent.

- **EVAPORATION:** Evaporators for material recovery and water reuse in the metal finishing industry are a proven technology with expanding applications. Evaporation is a simple concentration process. Water is evaporated from a solution until the chemicals remaining in the wastewater are concentrated to a level that allows their reuse in the process bath. Although evaporation has the highest energy requirements of the recycle alternatives, its simplicity and reliability are compensating factors.

❑ ELECTROLYTIC METAL RECOVERY

Electrolytic Metal Recovery (EMR) is being used successfully by electroplaters, rolling mills, printed circuit board manufacturers, and metal coating firms. A static rinse with an EMR unit is usually situated downstream of the process tank to remove the bulk dragout. Solution from the static rinse tank is circulated through an electrolytic cell where the metal is removed.

In the cell, a direct current is passed through the metal-bearing solution. Electromechanical reduction of metal ions to elemental metal takes place at the cathode. Simultaneously, oxygen is evolved at the anode.

Metal deposited on the cathode is allowed to build up to a thickness of about one-half inch. The power is then shut off and the deposited metal is recovered from the cathode. Modern electrolytic cells, under favorable conditions, can recover 99 percent of the dissolved metal in rinse solutions. Because the electrolytic process maintains a low concentration of metal in the dragout tank, the amount of metal carried into the succeeding rinse tanks is minimized. This, in turn, reduces the load to the downstream waste treatment plant reducing production of sludge.

A chief objection to EMR is the cost of power required to deposit metal from low concentration solutions. This is increasingly offset by the rising cost of primary metals.

❑ REVERSE OSMOSIS

In reverse osmosis, a waste solution is pumped under pressure into a chamber containing a semipermeable membrane. Only solvent (water) passes through the membrane, salts and dissolved metals remain. Two output streams are produced: one very concentrated with metals and one relatively clean. The clean water can be reused or discharged, and the concentrated solution can be used in the process bath. (See Figure M-7.)

Reverse osmosis has been successfully used on the rinse waters from a number of electroplating baths. However, chromic acid and high pH cyanide baths are not easily treated with reverse osmosis.

Reverse osmosis recovers plating solution additives in addition to dissolved metals. This is a major advantage. Unfortunately, reverse osmosis units are expensive to install, maintain, and operate.

ULTRAFILTRATION

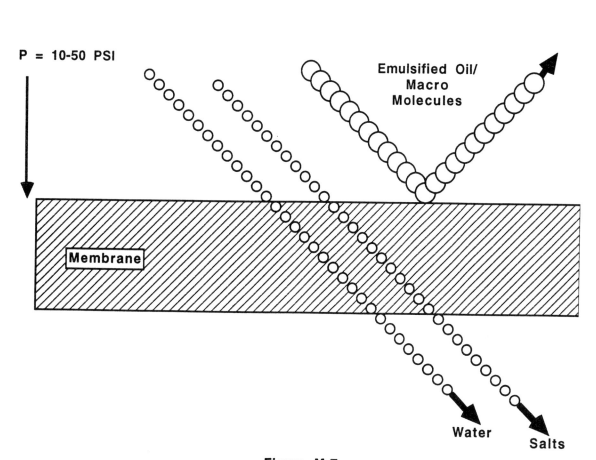

Figure M-7

❑ ION EXCHANGE

Ion exchange units consist of a resin bed or beds designed to remove cations (positively charged ions) or anions (negatively charged ions) from a waste stream. The liquid waste flows through exchange beds; hazardous ions from the waste bind to the bed material, displacing comparatively innocuous ions. Periodically, the resin beds must be chemically treated to remove the absorbed waste. For this reason, ion exchange units are usually installed in parallel; the waste stream is diverted through the second unit while the first one is being regenerated. The material removed from the resin beds often can be used in process baths. (See Figure M-8.)

Advantages of ion exchange are that water savings are significant, energy consumption is relatively low, and the process works well on dilute waste streams.

Disadvantages are high costs for chemicals, labor, and maintenance. In addition, capital costs are high and the process is not capable of producing a highly concentrated stream for recycling.

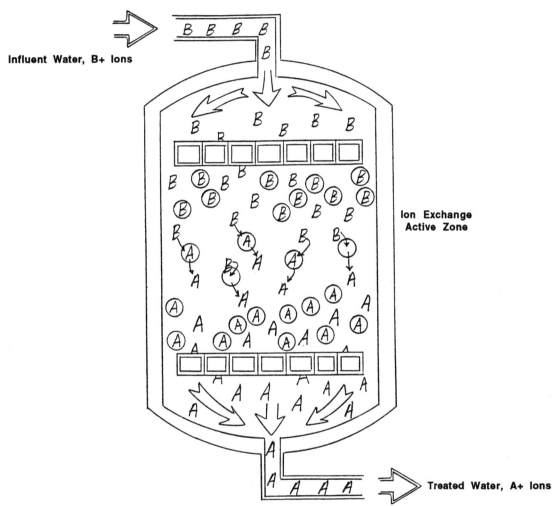

B+ ions displace A+ ions from the resin. The B+ ions remain bound to the resins. In waste treatment applications B+ ions are the hazardous constituents and A+ ions are comparatively innocuous.

ION EXCHANGE COLUMN

Figure M-8

❑ CRYSTALLIZATION

Heavy metals can be removed from saturated solutions with crystallization. This process, like EMR, removes a portion of the metal to form a solid that can be sold, reused, or dewatered and landfilled. Crystallization is best suited for on-site treatment waste in which stream segregation is possible and metal concentrations are high, such as in an etch or strip tank.

Commonly, a portion of the process tank solution is circulated through a refrigeration unit where crystallization takes place. The newly formed crystals are separated by gravity or filtration. With proper adjustment of the system, the desired metal concentration can be maintained in the process tank.

❑ ELECTRODIALYSIS

An electrodialysis apparatus is a series of alternating cation and anion permeable membranes between two electrodes. Wastewater passes between the parallel membranes. An electrical potential applied across the membranes causes the ions to migrate toward the electrodes. Alternate cells become either loaded with ions or depleted of them. The concentrated streams are sent to the process stream while the dilute stream goes to the rinse tank.

Electrodialysis has been shown to be an effective method for concentrating rinse waters to a high percentage of their original bath strength. The economic feasibility of the process depends heavily upon the life of the membranes which are susceptible to fouling and to leakage.

Advantages of electrodialysis includes low energy consumption, production of a highly concentrated stream for recovery, and good selectivity so that many undesirable impurities are eliminated.

❏ pH NEUTRALIZATION

Biological wastewater treatment proceeds optimally at a pH near 7 (neutrality). Small deviations from this value (+/- 1 unit) will reduce treatment efficiency, and large deviations may result in total inactivation of the bacteria. Also, pH's lower than five can seriously corrode concrete sewers and produce potentially deadly H_2S gas, while higher values (above 11) can cause calcification and occluding of sewer lines.

- **Batch**: Batch pH treatment simply means that contaminated wastewater is stored in a tank until such time that discharge is desired. The pH of a sample is determined and acid or caustic is added as needed. Tank contents are mixed, the pH retested and, if satisfactory, the contents are discharged. If not satisfactory, the process is repeated until the pH of the tank contents is within allowable limits.

- **Marble Chips**: The marble chip technique for pH neutralization consists of a control bed of limestone (calcium carbonate) chips through which acid wastewater flows continuously and is neutralized. These neutralization beds must be checked regularly as the limestone content can become fouled or exhausted.

- **Continuous pH Treatment System**: The most sophisticated pH treatment or controlling system is a continuous treatment system in which pH sensors, recording and control equipment, pumps and mixers are used to automatically control pH. In a continuous system, wastewater flows into a tank, a pH electrode senses the pH and sends a signal to a pump that adds the appropriate neutralizing chemical to the tank. An agitator then mixes it with the wastewater. The major variations for individual applications are the location of the electrodes and the mode of control - feed forward, feed backward, etc. One of the major considerations/problems with this otherwise excellent system is electrode maintenance. The electrodes foul easily and regular maintenance is mandatory to keep them operational (never longer than one week between maintenance periods; more frequently depending on the composition and character of the discharge effluent). A second, equally valid problem is inattention to regular standardization of the pH meters. If a probe or meter erroneously feeds back normal pH readings to the controller (when in actuality the pH is high or low), the system is not operational and the effluent is out of compliance - regardless of meter readings or strip chart recordings.

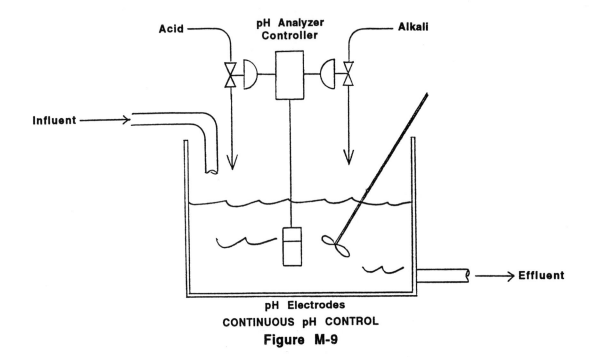

CONTINUOUS pH CONTROL
Figure M-9

❑ OXIDATION/OZONATION

Ozone (allotropic form of oxygen) is a good oxidizing agent and can be used to treat process wastewaters that contain cyanide. In fact, ozone oxidizes many cyanide complexes (for instance, iron and nickel complexes) that are not broken down by chlorine. Ozonation is primarily used to oxidize cyanide to cyanate. (See Cyanide Destruction.)

❏ EQUALIZATION

The flow rate of wastewater from an industrial treatment plant into a treatment system, or into a sewer main, is normally not steady, but is instead a function of variations in the production rate, frequency of batch processing and tank dumping. The concentrations of pollutants will also vary accordingly. High concentrations are called "slugs", or "slug flows." Many industries take advantage of equalization basins or holding tanks to average out these peaks. This blends other incoming wastewaters with the "slug flows" prior to initiating pretreatment or prior to discharging to the sewer. By so doing, the industry is able to stay within their discharge limits. There is no inference here that dilution (using uncontaminated water) can be introduced into this equalization equation. The purpose of an equalization procedure is to allow normally occurring process waters to be allowed to mix together in order to flatten out "highs and lows" that are an unavoidable occurrence in the industrial waste water process.

❑ GREASE AND OIL CONTROL SYSTEMS/DEVICES

Sewer systems with restaurants, fast-food establishments, large institutions and other more industrially oriented facilities which handle grease and oil (such as machine shops) are especially prone to sewer blockages. Extra care must be taken to ensure that excessive quantities of grease and oil are not discharged to the sewer system. Most sewer agencies have limitations for a maximum allowable discharge concentration of grease and oil. Discharge of grease and oil in excess of these limits can be a major cause of sewer line restrictions, blockages and wet well and pump station foulings. The blockages can also result in sewer system overflows which often lead to contaminated streams and beaches and the destruction of personal property. The wet well and pump station foulings can cause unnecessary time and money being spent by the sewer agency just to maintain these sites, not to mention the cost of investigating the cause of these grease and oil events.

If in-house control of grease and oil is insufficient, suitable grease and oil collection, or separation, or other control devices should be installed. Typical devices are shown on the following pages.

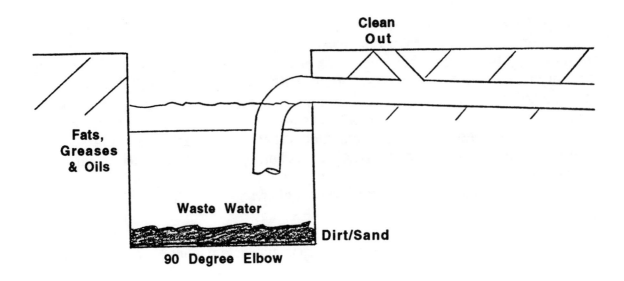
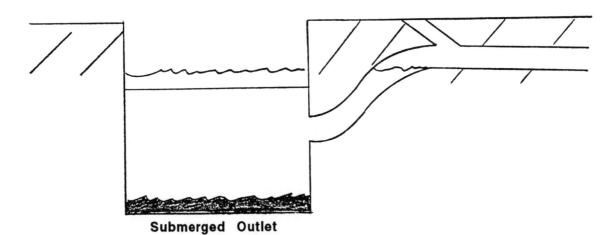

SINGLE CHAMBERED SUMPS
(For Low Volume Industrial Discharger; Ex: Steam Cleaning)
Figure M-10

M-23-a

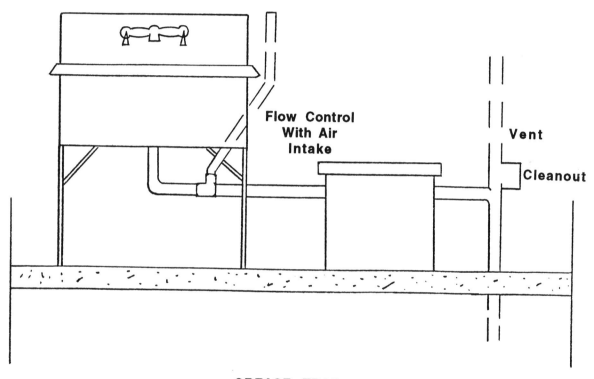

**GREASE TRAP
FOR A SMALL FOOD SERVICE ESTABLISHMENT
Figure M-11**

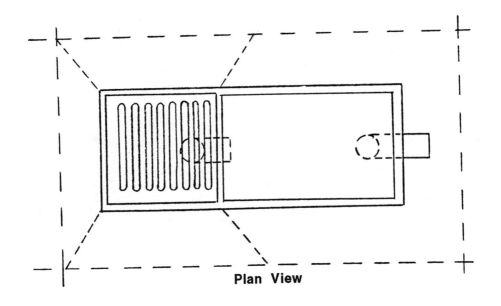

Plan View

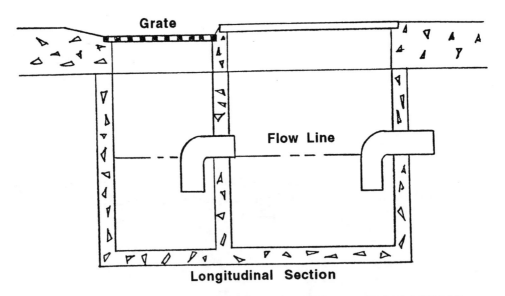

Longitudinal Section

TWO COMPARTMENT SAND AND GREASE INTERCEPTOR

Typically Used at a Car Wash Where the Unit is Inside the Wash Bay

Figure M-12

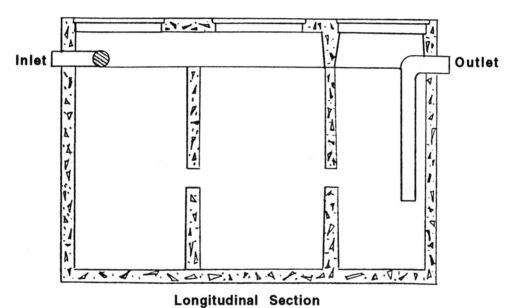

Longitudinal Section

TWO COMPARTMENT SAND AND GREASE INTERCEPTOR

Typically Used at a Car Wash Where the Unit is Outside the Wash Bay

Figure M-13

FOR HIGH VOLUME INDUSTRIAL DISCHARGER
(Example: Car Wash)

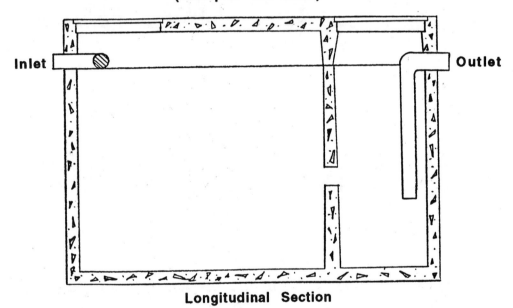

Longitudinal Section
THREE COMPARTMENT SAND AND GREASE INTERCEPTOR
Figure M-14

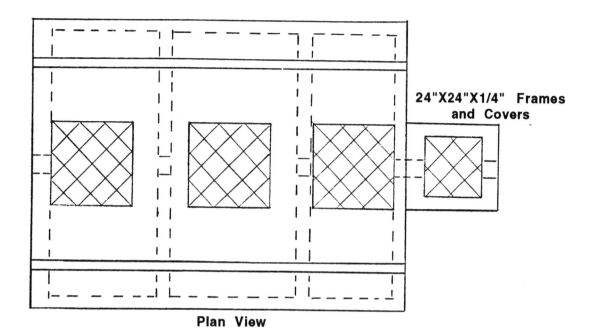

Plan View

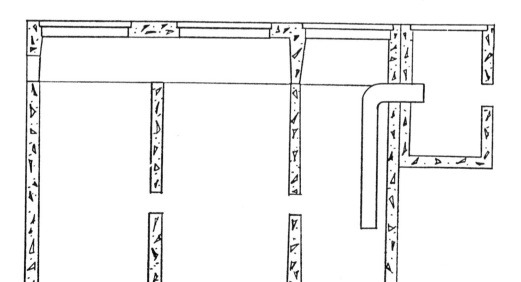

Longitudinal Section

**THREE COMPARTMENT INDUSTRIAL WASTE CLARIFIER
FOR HIGH VOLUME INDUSTRIAL DISCHARGER
(Example: Commercial Laundry or High Volume Restaurant)
Figure M-15**

❑ RINSING TECHNIQUES

The various types of rinsing techniques commonly used in the electroplating/metal finishing industry are described below:

- **Countercurrent Rinsing**: In counter current rinsing, the work piece is rinsed in several tanks in a series. Water flows counter to the movement of the work piece so that clean water enters the last rinse tank from which the clean product is removed. Wastewater is discharged from the first rinse tank which initially receives the contaminated product needing to be rinsed.

    Countercurrent rinsing provides for the <u>most</u> efficient water usage and, whenever possible, the countercurrent rinse method should be used. There is only one fresh water feed for the entire set of tanks and it is introduced into the last tank of the arrangement. The overflow from the last tank becomes the feed for the tank preceding it, and so on. Thus the concentration of dissolved salts decreases rapidly from the first to the last tank. (See Figure M-16.)

    In a situation requiring a 1,000 to 1 concentration reduction, the addition of a second rinse tank (with a countercurrent flow configuration) will reduce the theoretical water demand by 97 percent.

- **Closed Loop Rinsing**: Closed loop rinsing, as indicated in Figure M-18, is essentially the same as the countercurrent rinsing technique with one significant addition. Instead of the overflow from tank number one going to wastetreatment or some other control technology, the effluent goes to an evaporation/cooling tower system. The wastewater effluent is concentrated in the evaporator, the distillate condensed by a cooling system and returned to the rinse cycle (see Figure M-17).

- **Series Rinsing**: The major advantage of the series rinse over the countercurrent system is that the tanks of the series can be individually heated or level controlled since each has a separate feed. Each tank

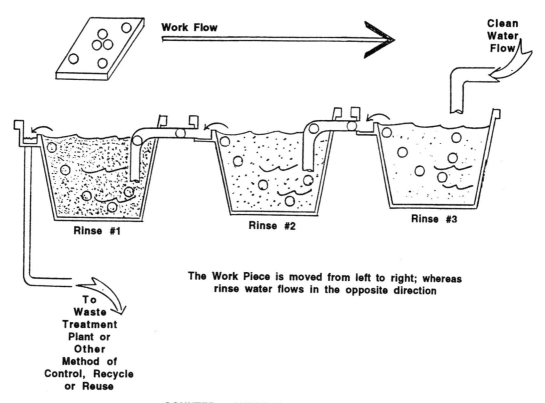

COUNTER - CURRENT RINSING
Figure M-16

M-24-a

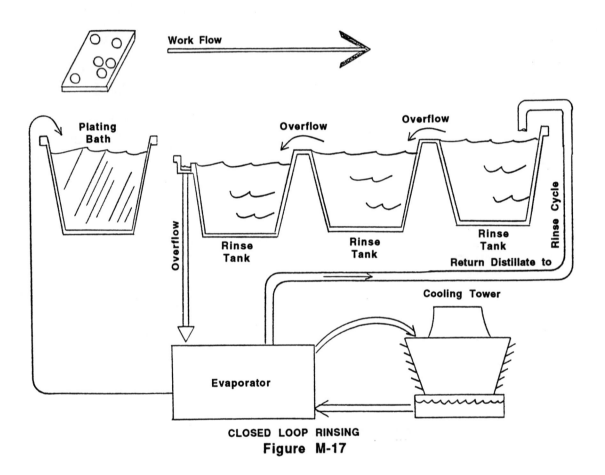

CLOSED LOOP RINSING
Figure M-17

M-24-b

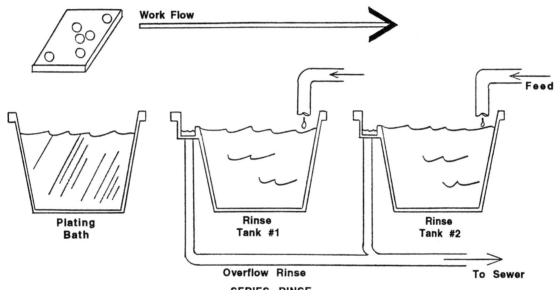

SERIES RINSE
Least Effective in Conserving Water

Figure M-18

reaches its own equilibrium conditions; the first having the highest concentration of dissolved salts, and the last rinse having the lowest concentration. This system uses water more efficiently than the single running rinse, but not as efficiently as the countercurrent system.

- **Single Running Rinse**: The single running rinse requires a large volume of water to effect an associated large amount of contaminant removal. Although in wide-spread use, when identified, a single running rinse should be modified or replaced by a more effective rinsing arrangement. The addition of even a single "still" or "dead" rinse tank prior to the single running rinse greatly reduces the contaminant level introduced into the rinse line.

- **Spray Rinse**: Spray rinsing is considered to be the _most_ efficient of the various rinsing techniques using continuous dilution rinsing. The main concern encountered in its use is the efficiency of the actual spray (i.e., the volume of water contacting the part and removing contamination compared to the volume of water discharged). This technique is very well suited to flat parts; the impact of the spray also provides an effective mechanism for removing dragout from recesses with a large "width to depth" ratio.

- **Dead, Still or Reclaim Rinses**: This form of rinsing is particularly applicable for initial rinsing after metal plating. The dead rinse allows for easier recovery of the metal and lower water usage. The rinse water can often be transferred to the plating tank that precedes it. The dead rinse is followed by spray or other running rinses. (See Figure M-19.)

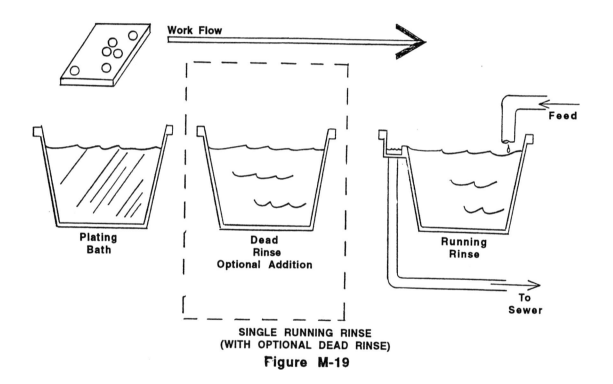

SINGLE RUNNING RINSE
(WITH OPTIONAL DEAD RINSE)
Figure M-19

- **Drag Out Reduction**

    Although technically speaking not really rinsing, drag out reduction techniques need to be mentioned as very useful pretreatment methods used by platers. By means of these techniques a plating firm can drastically reduce its rinse water consumption and pollutant loadings. Inspectors need to be aware of the following general techniques:

    Increase Parts Draining Time: Simply holding plated parts above the plating bath on a rack in order to allow solution clinging to the parts to drain back into the bath will reduce metals drag out. The longer the drainage time, the less drag out up to a point.

    Air Blowing: Some firms employ an air nozzle to speed drainage back into the bath. Excess solution is blown off parts thus reducing carry over to the rinses.

    Fog Rinsing: As mentioned above under spray rinses. In this case the amount of water rinse is deliberately minimized and the "fog" rinsing takes place over the plating bath so that the rinse falls back into the bath rather than entering the rinse drain. Fog or mist nozzles can be designed to effectively balance evaporation losses in the plating bath.

    Drag Out Additives: Some plating firms have experimented successfully with additives which reduce viscosity and/or surface tension of bath solutions, enhancing drainage back to reduce drag out. Wetting agents have been widely used. These techniques are usually nominal in their effect, i.e., "fine tuning."

❑ **Combination Rinsing**

In many plating shops where multiple lines are employed to plate a variety of metal ions, there exists the possibility for a variety of creative rinsing combinations (which can take advantage of specific process characteristics and requirements) to reduce water use and minimize metal pollutants.

The possible arrangements are quite numerous but all rely on the basic principles outlined in this section, from single plating lines to multiple lines incorporating mixed rinses. In most cases combination rinsing will be found in new plating plants where it was included in the original process design. Older multi-line shops are usually reluctant to undertake the restructuring needed to incorporate combination rinsing techniques.

All rinsing techniques point up the fact that it is often more productive for an existing plating operation to look at internal process modifications and process sequence changes than it is to provide end-of-pipe pretreatment without investigating such possibilities.

Theoretical Rinse Water Flows Required
to Maintain a 1,000 to 1 Concentration Reduction

Type of Rinse	Single	Series	Countercurrent
Number of Rinses	1	2 - 3	2 - 3
Required Flow (gpm)	10	0.61 - 0.37	0.31 - 0.10

Figure M-20

―――――― **Section N** ――――――

OIL RECOVERY AND OILY WASTE PRETREATMENT

Introduction

Many different types (or compounds) of oils and related fluids are common in oily wastes. These include cutting oils, machining oil, lubricants, greases, solvents, and hydraulic fluids.

The concentration of oil in wastewater is usually reported as Oil and Grease (O & G), or Fats, Oils and Grease (FOG). Oil is a broad category of hydrocarbon chemicals which encompasses gasoline, kerosine, crude oil, No. 6 distillate, bacon grease, and chicken fat, etc., all of which show up in the analysis via Freon extraction from water.

Oil can be present in wastewater in one or more of three distinct forms - and its form will directly affect its removability:

- FREE OIL - easily rises to the surface and can be readily skimmed off by any of several available mechanisms

- EMULSIFIED OIL - more difficult to remove because first the emulsion must be broken - certain additives can cause oil and water to mix thereby causing a most difficult waste treatment problem. (Laundry wastewater is an example.)

- DISSOLVED OIL - can be removed biologically or by adsorption with activated carbon

Removal of excessive amounts of waste oil from wastewater flows is crucial for sound operation of a sewerage system. If large amounts of oil are discharged to sewer systems and/or the treatment plant(s) several problems arise:

- Sewer mains can become occluded and wastewater flows are slowed down or even become obstructed causing sewer overflows - a very undesirable condition, especially in recreational beach areas.

- Screens become fouled and clogged in pump stations and treatment plants

- Skimming operations become more problematic at the treatment plant

- Wet wells and pump stations become fouled and require excessive maintenance and/or vacuuming and cleaning

- Float systems and electrodes become fouled, causing pumping system malfunctions

- Fire and explosion hazards are introduced into the entire system

- Biological activity in wastewater treatment may be inhibited at concentrations above 50 mg/l

By segregating oily wastes from other wastewaters, the expense of both the wastewater treatment and the oil recovery process is minimized, since the quantity and number of constituents involved are reduced. Also, segregated oily wastes are better suited for hauling to disposal or reclamation if on-site treatment is not a possibility. Additional segregation of oily wastes, by specific type or compound, can further reduce treatment or hauling costs as some oils have higher reclaim values, especially if not contaminated with other less desirable oils.

The areas that will be covered in this section are as follows:

❑ Types of Oil Recovery/TSDF's

❑ Pretreatment of Oily Wastes

❑ Inspector's Questions

❑ Types of Oil Recovery

Reprocessing of oil consists of contaminant removal by physical separation, filtering, centrifuging (De Lavaal centrifuge is an example) or magnetic separation. Reprocessing also includes the preparation of waste oils for burning as a fuel supplement.

Reclamation of oil combines the elements of reprocessing with mechanical or chemical steps. Reclamation removes solids, water, fuel or solvents and degradation products such as acid. Two common processes are Flash Distillation and Chemical Adsorption. Partial vacuum and filtration is then added to remove the degradation products from the used oil. Reclamation is used with synthetic fluids or highly refined mineral oils. Reclamation systems can be fixed or portable and outside reclamation services are also available.

Recycling of oil is, however, the most comprehensive treatment. The waste oil is prefiltered to remove the bulk of the solids, solvents/fuel and water. What remains is essentially base oil and additives. After removing the additives a high quality basestock remains. This basestock is then reformulated with new additives and can be used again as the virgin basestock. This method of re-refining provides the best economics when large volumes of waste oil are readily available.

Any of the procedures previously mentioned for removing oils from water are commonly employed at sites designed for the specific purpose of either (1) providing a service to the industrial community by accepting all types of mostly dilute, oily waste waters for treatment and disposal to the sewer (usually bilgewater) or (2) more concentrated waste oils are accepted primarily for recycling or re-refining, but treatment and disposal of low volumes of process wastewater could also occur as a result of this operation. In both cases, these processes would occur at a designated and permitted facility called a TRANSFER, STORAGE AND DISPOSAL FACILITY (TSDF). These sites are usually permitted by the State Department of Health Services (DOHS) and legally no site is allowed to transfer, store or dispose of off-site wastes without this TSDF permit even if, for example, they are able to adequately treat wastewater and meet sewer discharge limits. In theory a local wastewater discharge permit could be issued but the site would not be in full compliance without the associated TSDF permit. For that reason there is no validity to performing an industrial waste inspection until at least the Interim Status Document (ISD) has been processed by a site which intends to engage in this type of operation.

❑ Pretreatment of Oily Wastes

Gravity separation is the first or primary stage in oil removal from water. Simply stated, the wastewater is allowed to sit in a vessel, in a quiescent state, and the free oil, lighter than water, rises to the surface. Once the phase separation is completed, the top oil layer is belt or rope-skimmed or pumped from the surface to a waste oil tank.

Tanks utilized for this purpose, and of a particular design, are called API (American Petroleum Institute) Separators. The design of an API Separator is based on the gallons of wastewater/square foot of area/ minute. Also, the size of the oil droplets are in indirect proportion to the size of the separator and the time needed to bring about phase separation. The same is true for higher density oils.

Parallel Plate Interceptors (PPI) can also be used in conjunction with API Separators to enhance the gravity sedimentation/oil removal process. The PPI unit is composed of a series of parallel plates between which the wastewater flows. The plates cause an increased "calming action" on the wastewater and a reduction of the distance an oil droplet has to move vertically before being removed. The net effect of PPI's is increased efficiency and it can accomplish (1) the same percent removal in a smaller tank or (2) greater efficiency of separation in the same sized separator.

Emulsion breaking is usually brought about by a pH change or the addition of other chemicals such as lime or polyelectrolytes. Lime was the chemical of choice some years ago with polyelectrolytes now more popular because less sludge is produced using it. There is an optimum chemical dose at an optimum wastewater pH. Dose is important and too little or too much chemical can markedly effect efficiency. (See Figure N-2)

Air flotation is used after the emulsion has been successfully broken. Oil is normally removed by two different types of air flotation:

- Dissolved
- Induced

In dissolved air flotation (DAF) systems, the wastewater is pressurized and in

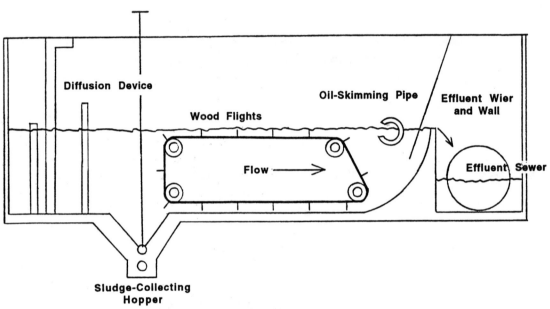

OIL/WATER SEPARATOR
Figure N-1

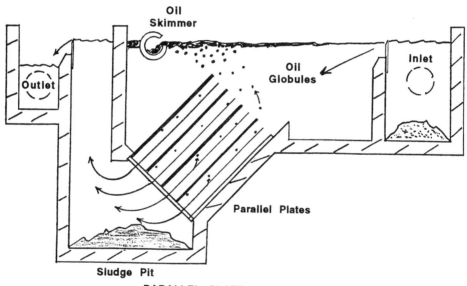

PARALLEL PLATE INTERCEPTOR
Figure N-2

the presence of added air, air dissolves and super saturates the water. When the pressure is released and the wastewater flows into an open flotation tank, small bubbles form and carry the free oil (and the suspended solids) to the surface where they can be removed.

Induced air flotation (IAF) causes bubbles to be formed in the water by means of drawing air beneath the surface of the liquid by either a high speed impeller rotating in a draft tube or through the throat of a nozzle venturi. Bubbles produced in this fashion are usually an order of magnitude larger than in dissolved air flotation. In IAF, bubble production and oil removal both take place in the same tank. (See Figure N-3.)

Ultrafiltration is also an alternative method for treatment of emulsified oil. In this process, oil-contaminated wastewater is pumped past a membrane. Under applied static pressure, water and most of the dissolved substances (chlorides, sulfates, etc.) flow through the pores of the membrane. The large molecules (emulsified oil being one of them) are retained. The permeate (water passing through the membrane) is relatively oil-free (less than 100 mg/l) but could still contain dissolved ions of consequence; heavy metals for example.

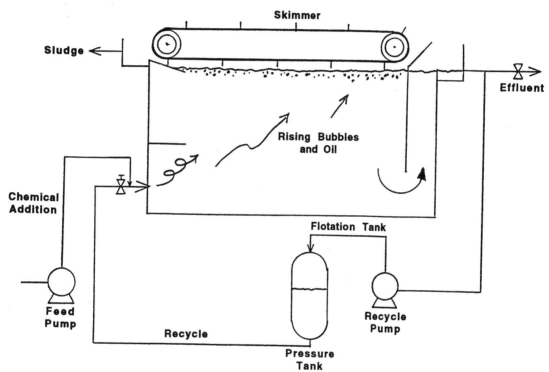

SIMPLIFIED SCHEMATIC OF AIR FLOTATION EQUIPMENT
Figure N-3

Results of Chemical Treatment of an Oily Wastewater

Parameter	Chemical Feed (mg/l)	Product (mg/l)
Oil and Grease	6036	76
Suspended Solids	1640	64
Lead	22	1.2
Zinc	12	1.1
pH	12.1	9.5

A major problem with chemical treatment is that every pound of chemical added produces about 20 pounds of sludge (5% solids and 95% water) which most likely has to be classified as hazardous waste. This imposes a second major cost for disposal as well as setting up a scenario for future disposal site liability.

Oil Removal Unit Operations

(Expected Results)

Stage Effluent	Process	Approximate Quality (mg/l)
Primary	API Separator	100
Secondary	Air Flotation	10
Tertiary	Activated Carbon	1

Figure N-4

Skimming Performance Data for Oil and Grease (mg/l)

Oil and Grease Influent (mg/l)	Type of Skimmer	Oil and Grease Effluent (mg/l)
395,538	API	13.3
53,800	API	16.0
19.4	Belt	8.3
61.0	Belt	14.0

Figure N-5

INSPECTOR'S QUESTIONS—OIL RECOVERY/OILY WASTES

1. What is the nature of the oil recovery operation at this site?
    - Reclamation?
    - Recycling?
    - Reprocessing?
    - Re-refining?
    - Some other type of treatment which removes or recovers waste oil?

2. Describe the treatment procedure of process used and include any waste streams produced as a result of this treatment. Identify the volume of wastewater produced and the discharge point...

3. What chemicals, if any, are used in the treatment?
    - Volumes used?

4. What are the individual volumes of the process tanks?

5. Is there any routine washdown of the work area(s)?
    - Where is the discharge point?

6. Are there any floor drains in the process area?
    - Does their flow pass through a common interceptor point?
    - If no, where does it discharge?

7. Do all process streams flow to a common interceptor point?
    - If yes, where is that point?
    - If no, where are the various points?

8. How is on-site generated hazardous wastes handled?
    - Who is the waste hauler?
    - What is the average volume of material disposed to landfill, recycled, etc.?
    - Where are the records kept? (Can ask to see manifests.)

9. Is any cooling water used at this facility? Explain...
   - Cooling towers - identify and evaluate losses and discharges.
   - Single-pass cooling water - identify and itemize daily average flows and discharge locations (single-pass usually not allowed by local sewer ordinance.)
   - Identify any cooling tower using chromates in water treatment and locate sample point(s)

10. Is there any water reuse in the plant?
    - If yes, where and what is it?

11. Any requirements for steam generation or use?
    - If yes, is make-up water metered?
    - Volume (gallons per day)?
    - Is there blowdown to sewer? ...Frequency? ...Volume (gallons per day)?
    - What are the steam losses (gallons per day)?

12. What is the nature and average gallons per day (gpd) of materials brought on-site for treatment?
    - Oily wastes? Describe...
    - Oily water? Bilge water? Describe...

13. Are the materials being treated, stored or disposed of in accordance with the TSD permit? Explain why or why not.

14. Have excess capacity fees either been paid for or excluded from payment? Explain why or why not.

15. Is there adequate spill containment in the process area(s)?
    - If yes, describe.
    - If no, describe.
    - How is the sewer otherwise protected from leaks or spills in the process area(s) and throughout the facility? Describe in detail.

Section O

CATHODE RAY TUBES (CRT)

Introduction

Electron tubes are devices in which electrons or ions are conducted between electrodes through a vacuum or ionized gas within a gas-tight envelope which may be glass, quartz, ceramic, or metal. A large variety of electron tubes are manufactured, including klystrons, magnetrons, cross field amplifiers, and modulators. These products are used in aircraft and missile guidance systems, weather radar, and specialized industrial applications. The Electron Tube subcategory includes cathode ray tubes and TV picture tubes that transform electrical current into visual images. Cathode-ray tubes generate images by focusing electrons onto a luminescent screen in a pattern controlled by the electrical field applied to the tube. In TV picture tubes, a stream of high-velocity electrons scans a luminescent screen. Variations in the electrical impulses applied to the tube cause changes in the intensity of the electron stream and generate the image on the screen.

Processes involved in the manufacture of electron tubes include degreasing of components; application of photoresist, graphite, and phosphors to glass panels; and sometimes electroplating operations including etching and machining. The application of phosphors is unique to TV picture tubes and other cathode-ray tubes. The phosphor materials may include sulphides of cadmium and zinc and yttrium and europium oxides. The electroplating operations are covered under the Metal Finishing Category. Raw materials can include copper and steel as basis materials, and copper, nickel, silver, gold, rhodium and chromium to be electroplated. Phosphors, graphite, and protective coatings containing toluene or silicates and solders of lead oxide may also be used. Process chemicals may include hydrofluoric, hydrochloric, sulfuric, and nitric acids for cleaning and conditioning of metal parts; and solvents such as methylene chloride, trichloroethylene, methanol, acetone, and polyvinyl alcohol.

Phosphorescent Coatings

Phosphorescent coatings are coatings of certain chemicals, such as calcium halophosphate and activated zinc sulfide, which emit light. Phosphorescent coatings are used for a variety of applications, including fluorescent lamps, high-pressure mercury vapor lamps, cathode ray and television tubes, lasers, instrument panels, postage stamps, laundry whiteners, and specialty paints. The most important fluorescent lamp coating is calcium halophosphate phosphor. The intermediate powders are calcium phosphate and calcium fluoride. There are three TV powders: red, blue, and green. The red phosphor is yttrium oxide activated with europium; the blue phosphor is zinc sulfide activated with silver, and the green phosphor is zinc-cadmium sulfide activated with copper. The major process steps in producing phosphorescent coatings are reacting, milling, and firing the raw materials; recrystallizing raw materials, if necessary; and washing, filtering, and drying the intermediate and final products.

Electron Tubes Subcategory

The agency has insufficient information to adequately characterize pollutants from this subcategory. The major pollutants are identified as fluoride and lead.

Phosphorescent Coatings Subcategory

Data presently available to the agency are insufficient to adequately characterize the wastewater discharges for the Phosphorescent Coatings subcategory. However, major pollutants are suspended solids, fluoride, cadmium, and zinc.

Further inclusions are:

❑ PSES/PSNS Limitations

❑ Inspector's Questions

Table O-1

CATHODE RAY TUBE
PSES LIMITATIONS

POLLUTANT OR POLLUTANT PARAMETER	DAILY MAXIMUM	MAXIMUM MONTHLY AVERAGE
Cadmium	0.06	0.03
Chromium	.65	0.30
Lead	1.12	0.41
Fluoride	35.0	18.0
Zinc	1.38	.56
TTO	1.58	NA

The term "cathode ray tubes" means electronic devices in which electrons focus through a vacuum to generate a controlled image on a luminescent surface. This definition does not include receiving and transmitting tubes.

PSES Standards apply if: Facility construction or operation prior to 9 March 1983

Table O-2

CATHODE RAY TUBE
PSNS LIMITATIONS

POLLUTANT OR POLLUTANT PARAMETER	DAILY MAXIMUM	MAXIMUM MONTHLY AVERAGE
Cadmium	0.06	0.03
Chromium	.56	0.26
Lead	.72	0.27
Fluoride	35.0	18.0
Zinc	.80	.33
TTO	1.58	NA

The term "cathode ray tubes" means electronic devices in which electrons focus through a vacuum to generate a controlled image on a luminescent surface. This definition does not include receiving and transmitting tubes.

PSNS Standards apply if: Facility construction or operation after 9 March 1983

INSPECTOR'S QUESTIONS—CATHODE RAY TUBES

1. When did Cathode Ray Tube production begin at this facility? Day/Month/Year.

2. Is this facility regulated under CRT PSES or PSNS limitations?

3. What metals are used in solution or otherwise present in process tanks?

4. What, if any, metals are being plated? Explain...

5. Any etching being performed?
    - If yes, describe and note what chemical, i.e., acid
    - Where is the discharge point? Describe...

6. Is any lead soldering being done?
    - If yes, describe the operation and note the discharge point

7. Is any vapor degreasing done?
    - If yes, is the unit water cooled (one-pass), on a cooling tower or refrigerant type?
    - If one-pass, where does it discharge to sewer? (One-pass not usually allowed by sewer ordinance)
    - What type of solvent is used in the vapor degreaser?
    - How is the spent or contaminated solvent handled? Disposed of? Describe

8. What are the individual volumes and locations of the process tanks?

9. How are spent or contaminated processing tanks handled?
    - If hauled, who hauls it? (Can check manifests.)

10. Are any tanks heated?
    - If yes, which ones?

11. Any rectifiers, compressors or similar equipment in use?
    - If yes, do any of them require cooling water?

- Any one-pass cooling water?
- If yes, volume (gallons per day)?
- Where is the discharge point?
- If water is not used for cooling equipment, describe the cooling practices.

12. Is there any routine washdown of the work area(s)?
    - Where is the discharge point?

13. Are there any floor drains in the process area?
    - Does their flow pass through a common interceptor point?
    - If no, where does it discharge?

14. Do all process streams flow to a common interceptor point?
    - If yes, where is that point?
    - If no, where are the various points?
    - Is one-pass or brine reject water co-mingling with categorically regulated wastewater at the sample point(s)?
    - (See Combined Waste Stream Formula)

15. Is there any pretreatment of waste water?
    - If yes, describe the various aspects in place and where located.
    - If no, make recommendation or appropriate comments.

16. How is the disposal of hazardous wastes handled?
    - Who is the waste hauler?
    - What is the average volume of material disposed to landfill, recycled, etc.?
    - Where are the records kept? (Can ask to see manifests.)

17. Is there any water reuse in the facility?
    - If yes, where and what is it?

18. Is there any reverse osmosis or de-ionized water production?
    - If yes, is it on a service? Who maintains the equipment?
    - What, if any, pretreatment is there for the systems? Describe...
    - Is water kept in a holding tank and drawn off as needed?
    - Or is it produced on an as-needed basis?
    - Is the water metered in? Out?
    - What is the reject ratio for the reverse osmosis system; i.e., what percentage

of water goes back to the sewer as reject and what percentage is used as process water?
- Where is the sample point for the wastewater discharge?
- Does regulated wastewater from the cathode ray tube operation also flow to this location?

19. What types of rinse tanks, rinsing procedures or configurations are in use at this facility? Explain or describe...
    - Running rinses
    - Still rinses
    - Countercurrent rinses
    - Spray rinses
    - Fog rinses
    - Other

20. What solvents are used at this facility?
    - Where are they used? What unit operations?
    - What is the method of application? Dip tank, rag applied, or other?
    - Is there any discharge of solvents to sewer in any form? From the unit operation(s)? Final triple rinsing of containers prior to disposal? Explain...
    - If no solvents are discharged to sewer, explain what procedures are in place to prevent discharge. Solvent Management Plan (SMP)?
    - When was SMP submitted? Explain...

21. Any requirements for steam generation or use?
    - If yes, is make-up water metered?
    - Volume (gallons per day)?
    - Is there blowdown to sewer? ...Frequency? ...Volume (gallons per day)?
    - What are the steam losses (gallons per day)?

22. Is there adequate spill containment in the processing area(s)?
    - If yes, describe.
    - If no, describe.
    - How is the sewer otherwise protected from leaks or spills in the process area(s) and throughout the facility? Describe in detail.

Section P

SEMICONDUCTORS

Introduction

Semiconductors are solid state electrical devices which perform a variety of functions in electronic circuits. These functions include information processing and display, power handling, data storage, signal conditioning, and the interconversion between light energy and electrical energy. The semiconductors range from the simple diode, commonly used as an alternating current rectifier, to the integrated circuit which may have the equivalent of 250,000 active components in a 0.635 cm (1/4 inch) square.

Semiconductors are used throughout the electronics industry. The major semiconductor products are:

- Silicon based integrated circuits which include bipolar, MOS (metal oxide silicon), and digital and analog devices. Integrated circuits are used in a wide variety of commercial and consumer electronic equipment, calculators, electronic games and toys, and medical equipment.

- Light emitting diodes (LED) which are produced from gallium arsenide and gallium phosphide wafers. These devices are commonly used as information displays in electronic games, watches and calculators.

- Diodes and transistors which are produced from silicon or germanium wafers. These devices are used as active components in electronic circuits which rectify, amplify or condition electronic signals.

- Liquid crystal display (LCD) devices which are produced from liquid crystals. These devices are primarily used for information displays as an alternative to LEDs.

Silicone based integrated circuits will be discussed more fully in the following pages, to include:

❑ Silicone Based Integrated Circuits

- Circuits Production Flow Diagram
- PSES/PSNS Limits for Total Toxic Organics
- Inspector's Questions

❑ SILICON-BASED INTEGRATED CIRCUITS: Manufacturing Processes and materials

These circuits require high purity single crystal silicon as a basis material. Most of the companies involved in silicon-based integrated circuit production purchase single crystal silicon ingots rather than grow their own crystals.

When the ingot is received it is sliced into round wafers approximately 0.76 mm (0.030 inches) thick. These slices are then lapped or polished by means of a mechanical grinding machine or are chemically etched to provide a smooth surface and remove surface oxides and contaminants. Commonly used etch solutions are hydrofluoric acid or hydrofluoric-nitric acid mixtures. Other acids such as sulfuric or nitric may be used depending on the nature of the material to be removed. Wastewater results from cooling the diamond tipped saws used for slicing, from spent etch solution, and from de-ionized (DI) water rinses following chemical etching and milling operations.

The next step in the process depends on the type of integrated circuit device being produced, but commonly involves the deposition of growth of a layer or layers of silicon dioxide, silicon nitride or apitaxial silicon.

The wafer is then coated with a photoresist, a photosensitive emulsion. The wafer is next exposed to ultraviolet light using glass photomasks that allow the light to strike only selected areas. After exposure to ultraviolet light, unexposed resist is removed from the wafer, usually in a DI water rinse. This allows selective etching of the wafer. The wafer is then visually inspected under a microscope and etched in a solution containing hydrofluoric acid (HF). The etchant produces depressions, called holes or windows, where the diffusion of dopants later occurs. Dopants are impurities such as boron, phosphorus and other specific metals. These impurities eventually form circuits through which electrical impulses can be transmitted. The wafer is then rinsed in an acid or solvent solution to remove the remainder of the hardened photoresist material.

Diffusion of dopants is generally a vapor phase process in which the dopant, in the form of a gas, is injected into a furnace containing the wafers. Gaseous phosphine and boron trifluoride are common sources for phosphorus and boron dopants. The gaseous compound breaks down into elemental phosphorus or boron on the hot wafer surface. Solid forms of the dopant may also be used. For example, boron oxide wafers can be introduced into the furnace in close proximity to the silicon wafers. The boron oxide sublimes and deposits boron on the surface of the wafer by condensation and then diffuses into the wafer upon continued heating.

Finally, a second oxide layer is grown on the wafer, and the process is repeated.

During the photolithographic-etching-diffusion-oxide processes, the wafer may be cleaned many times in mild acid or alkali solutions followed by DI water rinses and solvent drying with acetone of isopropyl alcohol. This is necessary to maintain wafer cleanliness.

After the diffusion processes are completed, a layer of metal is deposited onto the surface of the wafer to provide contact points for final assembly. The metals used for this purpose include aluminum, copper, chromium, gold, nickel, platinum, and silver. The processes associated with the application of the metal layer are covered by regulations for the Metal Finishing category. One of the following three processes is used to deposit this metal layer:

- Sputtering: In this process the source metal and the target wafer are electrically charged, as the cathode and anode, respectively, in a partially evacuated chamber. The electric field ionized the gas in the chamber and these ions bombard the source metal cathode, ejecting metal which deposits on the wafer surface.

- Vacuum Deposition: In this process the source metal is heated in a high vacuum chamber be resistance or electron beam heating to the vaporization temperature. The vaporized metal condenses on the surface of the silicon wafer.

- Electroplating: In this process the source metal is electrochemically deposited on the target wafer by immersion in an electroplating solution and the application of an electrical current.

Finally, the wafer receives a protective oxide layer (passivation) coating before being back lapped to produce a wafer of the desired thickness. Then the individual chips are diced from the wafer and are assembled in lead frames for use. Many companies involved in semiconductor production send completed wafers to overseas facilities where dicing and assembly operations are less costly as a result of the amount of hand labor necessary to inspect and assemble finished products.

For an overview of silicon integrated circuit production, please see Figure LLL-1 on page LLL-5-a.

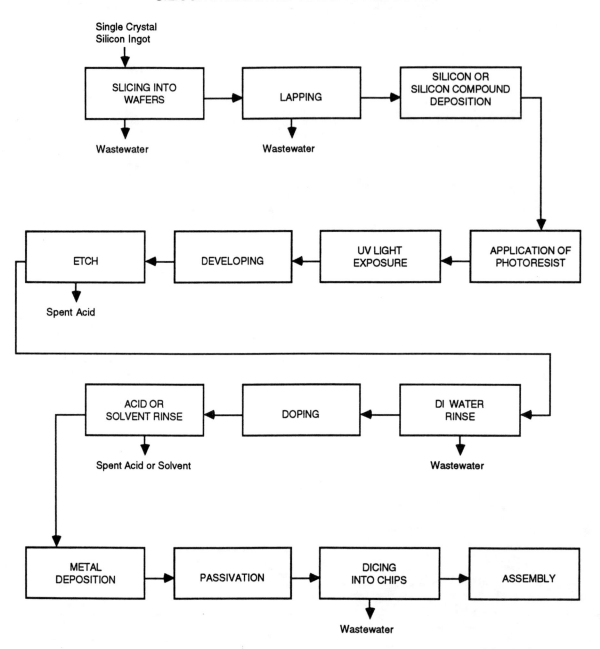

Figure P-1

Table P-1

SEMICONDUCTORS
PSES LIMITATIONS

POLLUTANT OR POLLUTANT PARAMETER	DAILY MAXIMUM	MAXIMUM MONTHLY AVERAGE
TTO	1.37	Not Applicable

PSES Standards apply if: Facility construction or operation date prior to: 24 August 1982

INSPECTOR'S QUESTIONS—SEMICONDUCTORS

1. When did Semiconductor production begin at this facility? Day/month/year.

2. What metals are in use, in solution or otherwise present in process tanks?

3. What, if any, metals are being plated?

4. Any etching being performed? Is any passivating being performed?
    - If yes, describe and note what chemical, i.e., acid is being used
    - Where is the discharge point

5. Any photo resist?
    - If yes, describe process and include chemicals used.
    - Where is the discharge point?

6. Are any solvents used in the process?
    - If yes, which solvents?
    - Any solvents discharged to sewer or do any find their way into the wastewater? ...Describe?
    - Where is the discharge point?
    - Solvent Management Plan? Date submitted?

7. Is any vapor degreasing done?
    - If yes, is the unit water cooled (one-pass), on a cooling tower or refrigerant type?
    - If one-pass, where does it discharge to sewer? What is the flow rate? (One-pass not usually allowed by local sewer ordinance.)
    - What type of solvent is used in the vapor degreaser?
    - How is the spent or contaminated solvent handled? Disposed of? Describe

8. How are spent or contaminated processing tanks handled?
    - If hauled, who hauls it? (Can check manifests.)

9. Are any tanks heated?
    - If yes, which ones?
    - What are the individual volumes and locations of all process tanks in facility - plating, acid, etc.?

10. Are alkaline or acid cleaning tanks present?
    - If yes, how frequently are these tanks batch discharged to the sewer?
    - What pretreatment occurs prior to discharge?
    - What are the individual volumes in all process tanks in facility - plating, acid, etc.?
    - How is bottom sludge handled?

11. Any rectifiers, compressors or similar equipment in use?
    - If yes, do any of them require cooling water?
    - Any one-pass cooling water?
    - If yes, volume (gallons per day)?
    - Where is the discharge point?
    - If water is not used for cooling equipment, describe the cooling practices.

12. Is there any routine washdown of the work area(s)?
    - Where is the discharge point?

13. Are there any floor drains in the process area?
    - Does their flow pass through a common interceptor point?
    - If no, where does it discharge?

14. Do all process streams flow to a common interceptor point?
    - If yes, where is that point?
    - If no, where are the various points?
    - Is one-pass or brine reject water co-mingling with categorically regulated wastewater at the sample point(s)?
    - (See Combined Waste Stream Formula.)

15. Is there any pretreatment of waste water?
    - If yes, describe the various aspects in place and where located.

16. How is the routine disposal of hazardous wastes handled?

- Who is the waste hauler?
- What is the average volume of material disposed to landfill, recycled, etc.?
- Where are the records kept? (Can ask to see manifests.)

17. Is there any water reuse in the plant?
    - If yes, where and what is it? Describe...

18. Is there any reverse osmosis or de-ionized water production?
    - If yes, is it on a service? Who maintains the equipment?
    - What, if any, pretreatment is there for the systems?...Describe?
    - Is water kept in a holding tank and drawn off as needed?
    - Or is it produced on an as-needed basis?
    - Is the water metered in? Out?
    - What is the reject ratio for the reverse osmosis system; i.e., what percentage of water goes back to the sewer as reject and what percentage is used as process water?
    - Where is the sample point for the wastewater discharge?
    - Does regulated wastewater from the Semiconductor operation also flow to this location?

19. What types of rinse tanks, rinsing procedures or configurations are in use at this facility? Explain or describe...
    - Running rinses
    - Still rinses
    - Countercurrent rinses
    - Spray rinses
    - Fog rinses
    - Other

20. Is there adequate spill containment in the processing area(s)?
    - If yes, describe.
    - If no, describe.
    - How is the sewer otherwise protected from leaks or spills in the process area(s) and throughout the facility? Describe in detail.

Section Q

COOLING SYSTEMS

Introduction

Cooling water treatment systems fall into three basic types:

- Open Recirculating Systems

- Closed Recirculating Systems

- Once-through (One-pass) Cooling

Open recirculating systems are of primary interest because they require fresh water make-up to replace losses from bleed (discharge to sewer) and/or evaporative losses. The function of bleeding is to rid the system of excess salts due to evaporative losses and solids concentration. This solids removal may be automatically controlled or done manually. Closed recirculating systems only require identification since water losses are normally non-existent and therefore do not affect the water audit. In general one-pass cooling systems are not usually allowed by the sewer agency because they introduce significant quantities of clean water into the sewer, thereby taking up unnecessary hydraulic capacity. The addition of clean water to the sewerage system is an unnecessary burden on treatment and transport and the waste of a valuable resource. Open or closed recirculating systems, or some combination of the two, is the preferred cooling system for most applications where reduction of temperatures is a consideration.

❑ Open Recirculating Systems

An open recirculating system incorporates a cooling tower or evaporation pond to dissipate the heat it removes from the process or product. An open recirculating system takes water from a cooling tower basin or pond, passes it through process equipment requiring cooling, then returns the water through the evaporation unit. The transfer of heat (equal to the ΔSH_{vap}) from the liquid to vapor phase lowers the temperature of the water that remains. Figure AA-1 shows a simplified open recirculating system as used in conjunction with a closed recirculating system and a heat exchanger.

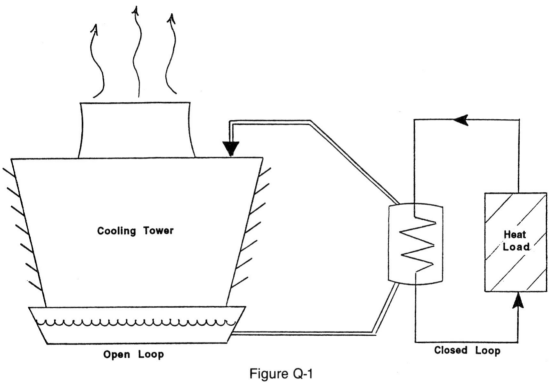

Figure Q-1

❏ Closed Recirculating Systems

A closed recirculating system is one in which the water is circulated in a closed loop with little evaporation or exposure to the atmosphere or other influences that would affect the chemistry of the water in the system. These usually require high chemical treatment levels to control biological activity (e.g., algae growth), and since water losses are negligible these levels are economical.

Heat is transferred to the closed cooling water loop by typical heat exchange equipment and is removed from the closed system loop by a secondary exchange of heat from the closed loop to a secondary cooling water cycle. Figure AA-1 shows a simplified closed recirculating system as used in conjunction with a closed recirculating system and a heat exchanger.

❑ Once-Through (One-Pass) Cooling

Once-through cooling systems take water from the plant supply, pass it through the cooling system and the object to be cooled and return it to the sewer. In this way heat is picked up from the object requiring cooling. One of the negative characteristics of a single-pass cooling system is the relatively large quantity of water required for cooling and the general prohibition against its discharge into the public sewer. In most applications of heat transfer, water can be circulated in a closed loop system and the less desirable option of single-pass cooling water discharge thereby eliminated. Figure AA-2 shows a simple flow diagram for a one-pass cooling water system.

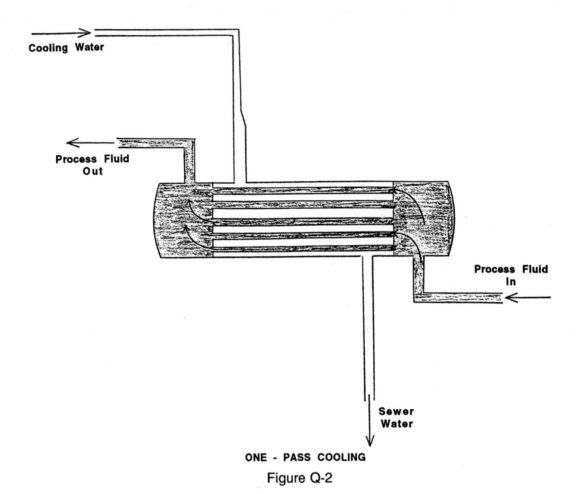

ONE - PASS COOLING
Figure Q-2

Cooling Towers Cooling towers are designed to evaporate water by intimate contact of water with air. Cooling towers are classified by the method used to induce air flow (natural or mechanical draft) and by the direction of air flow (either counterflow or crossflow relative to the downward flow of water). Figures AA-3 through AA-5 show examples of typical cooling towers.

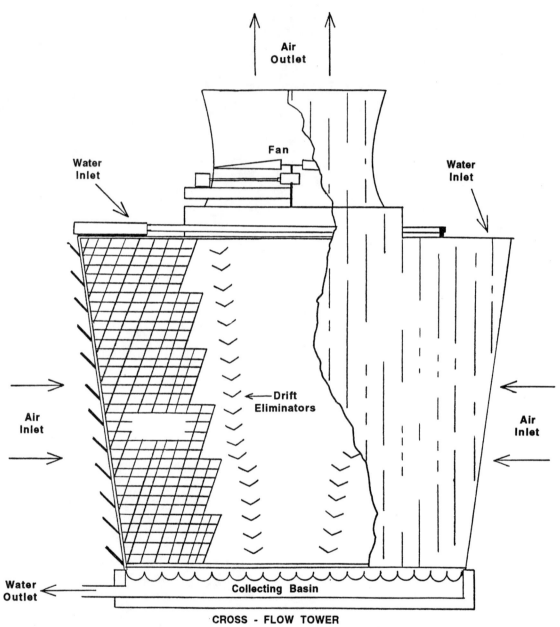

Figure Q-3 — CROSS-FLOW TOWER

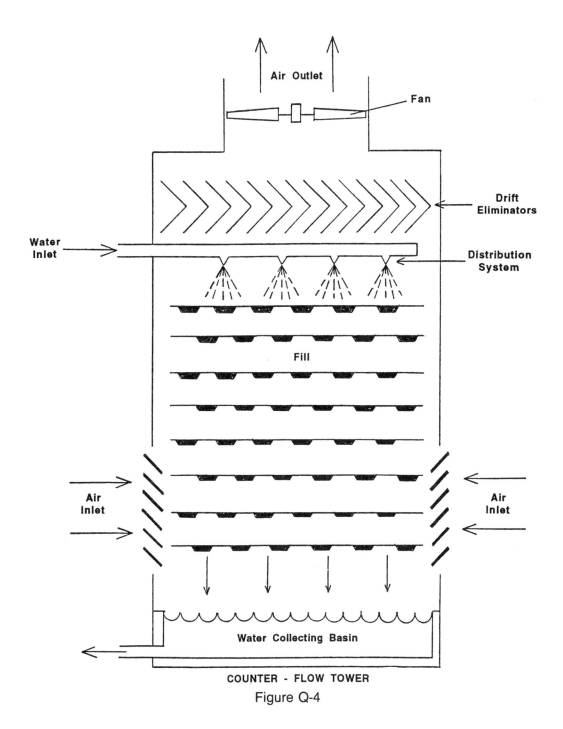

COUNTER - FLOW TOWER
Figure Q-4

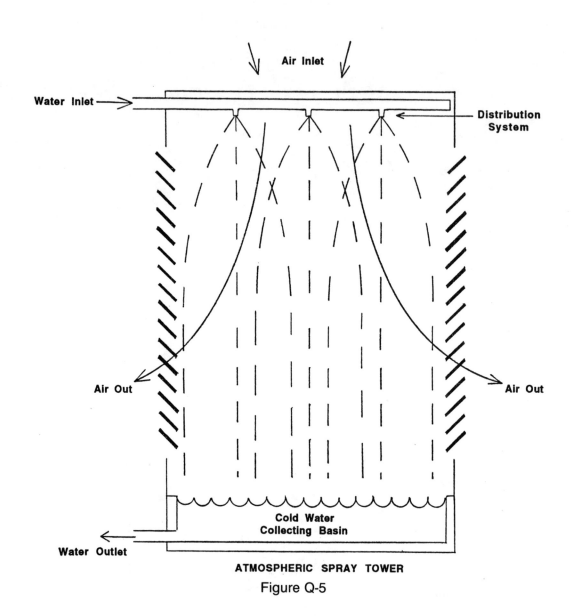

ATMOSPHERIC SPRAY TOWER
Figure Q-5

Cooling towers are typically found wherever water is used to cool a process or product. They are used in air conditioning, refrigeration, and other cooling applications. Look for cooling towers at:

- Laundries
- Packing Houses/Slaughter Houses
- Grocery Stores
- Cold Storage Facilities
- Dairies
- Ice Cream Storage Areas
- Dry Cleaners
- Welding/Machine Shops
- Hospitals (Where autoclaves are used)
- High Rise and Other Large Buildings..

INSPECTOR'S QUESTIONS—COOLING SYSTEMS

1. Determine if any type of cooling water system is used
    - Identify and describe including:
    - Volume of water used?
    - Tonnage of equipment?
    - Location of equipment?
    - Bleed volumes and frequencies?
    - Discharge points?
    - Chemicals used, especially chromates?

Bibliography

Black and Veach, 1981. *Final Report on the Development of the Metropolitan Sewerage System Industrial Waste Program.* Volumes 1 - 8: Kansas City, MO

The Bureau of National Affairs. 1988. *Water Pollution Control BNA Policy and Practice Series.* Washington, D. C.: The Bureau of National Affairs

Ecology and Environment, Inc. 1988. *Waste Audit Study General Medical and Surgical Hospitals.* California Department of Health Services

Eastman Kodak Company. 1986. *Environment J-20*: Rochester, NY

Eastman Kodak Company. 1987. *Environment J-21*: Rochester, NY

Effluent Guidelines Division, Office of Water Regulations and Standards, 1984. *Guidance Manual for Electroplating and Metal Finishing Pretreatment Standards.* Washington, D.C.: U. S. Environmental Protection Agency

EPA, Effluent Guidelines Division. 1983. (Final) *Development Document for Effluent Limitations Guidelines and Standards for the Metal Finishing Point Source Category.* Washington, D.C.: United States Environmental Protection Agency

EPA, Effluent Guidelines Division. 1982. (Proposed) *Development Document for Effluent Limitations Guidelines and Standards for the Pharmaceutical Point Source Category.* Washington, D.C.: United States Environmental Protection Agency

EPA, Effluent Guidelines Division. 1984. *Final Development Document for Effluent Limitations Guidelines and Standards for the Electrical and Electronic Components Point Source Category. Phase II.* Washington, D.C.: United States Environmental Protection Agency

EPA, Effluent Guidelines Division. 1981. *Guidance Document for the Control of Water Pollution in the Photographic Processing Industry.* Washington, D.C.: United States Environmental Protection Agency

Frank and Kemmer, ed. 1979. *The Nalco Water Handbook.* New York: McGraw Hill

HTM Division, Jacobs Engineering Group, Inc. 1987. *Waste Audit Study Paint Manufacturing Industry.* California Department of Health Services

Jacobs Engineering Group, Inc. Hazardous and Toxic Materials Division. 1988. *Waste Audit Study Commercial Printing Industry.* California Department of Health Services

Kushner and Kushner, 1981. *Water and Waste Control For the Plating Shop; 2nd Edition.* Gardner Publications, Inc.

March, April, May 1983. *Industrial Launderer*

McGraw-Hill Book Company, 1984. *Perry's Chemical Engineer's Handbook, 6th Edition.* New York, New York

Nemerow, Nelson L., 1978. *Industrial Water Pollution.* Addison-Wesley Publishing Company: Reading, MA

Potter, et. al, ed. 1986. *Alternative Technology for Recycling and Treatment of Hazardous Wastes.* Sacramento: State of California

Ralph Stone and Co., Inc., 1988. *Waste Audit Study - Research and Educational Institutions.* California Department of Health Services

Science Applications International Corp., ed. 1985. *State of California Pretreatment Program Implementation Guidance.* McLean, VA: SAIC

Conference Materials. University of Toledo, Division of Continuing Education. 1987. (Unpublished.) *Industrial Wastewater Pretreatment Water Conservation/Product Recovery/Pollution Abatement.* Toledo, OH: University of Toledo

Wesley, M., Toy, P. E., 1987. *Waste Audit Study Automotive Repairs.* California Department of Health Services

Glossary

Absorb - To take up matter or radiation.

Acid Rinse - A solution, usually dilute acetic acid, used as a stop bath following development.

Act - Federal Water Pollution Control Act.

Activate - To treat the cathode or target of an electron tube in order to create or increase the emission of electrons.

Activation - The process of treating a substance by heat, radiation or the presence of another substance so that the first mentioned substance will undergo chemical or physical change more rapidly or completely.

Additive Circuitry - 1. Full - Circuitry produced by the buildup of an electroless copper pattern upon an unclad board. 2. Semi - Circuitry produced by the selective "quick" etch of an electroless layer; this copper layer was previously deposited on an unclad board.

Adjustable Capacitor - A device capable of holding an electrical charge at any one of several discrete values.

Adsorption - The adhesion of an extremely thin layer of molecules (as of gas, solids or liquids) to the surface of solid or liquids with which they are in contact.

Adsorption - The adhesion of an extremely thin layer of molecules (of gas, liquid) to the surface of solids (granular activated carbon for instance) or liquids with which they are in contact.

Aeration - The addition of air to a liquid. This is done by pumping the liquid into the air or by bubbling air through it via sparging tubes. Aeration is used as part of the ferric EDTA bleach regeneration process in photographic processing. It can be used for reduction of oxygen demand in wastewater.

Aerobic - Living, active, or occurring only in the presence of oxygen.

Aerobic Biological Oxidation - Any waste treatment process utilizing organisms in the presence of air or oxygen to reduce the pollution load or oxygen demand of organic substance in water.

Aerobic Digestion - (Sludge Processing) The biochemical decomposition of organic matter, by organisms living or active only in the presence of oxygen, which results in the formation of mineral and simpler organic compounds.

Aldehydes Group - A group of various highly reactive compounds typified by actaldehyde and characterized by the group CHO.

Algicide - Chemicals used to retard the growth of phytoplankton (algae) in bodies of water.

Algicides - Chemicals for preventing the growth of algae.

Alkaline Cleaning - A process for cleaning basis material where mineral and animal fats and oils must be removed from the surface. Solutions at high temperatures containing caustic soda, ash, alkaline silicates and alkaline phosphates are commonly used.

Alkalinity - The capacity of water to neutralize acids, a property imparted by the water's content of carbonates, bicarbonates, hydroxides, and occasionally borates, silicates, and phosphates.

Alloy Steels - Steels with carbon content between 0.1% to 1.1% and containing elements such as nickel, chromium, molybdenum and vanadium. (The total of all such alloying elements in these type steels is usually less than 5%.)

Aluminizing - Forming an aluminum or aluminum alloy coating on a metal by hot dipping, hot spraying or diffusion.

Aluminum Foil - Aluminum in the form of a sheet of thickness not exceeding 0.005 inch.

Amines - A class of organic compounds of nitrogen that may be considered as derived from ammonia (NH_3) by replacing one or more of the hydrogen atoms by organic radicals, such as CH_3 or C_6H_5, as in methylamine and aniline. The former is a gas at ordinary temperature and pressure, but other amines are liquids or solids. All amines are basic in nature and usually combine readily with hydrochloric or other strong acids to form salts.

Anaerobic Biological Treatment - Any waste treatment process utilizing anaerobic or facultative organisms in the absence of air to reduce the organic matter in water.

Anaerobic Digestion - The process of allowing sludges to decompose naturally in heated tanks without a supply of oxygen.

Anaerobic Waste Treatment - (Sludge Processing) Waste stabilization brought about through the action of microorganisms in the absence of air or elemental oxygen.

Anhydrous - Containing no water.

Anions - The negatively charged ions in solution, e.g., hydroxyl.

Anneal - To treat a metal, alloy, or glass by a process of heating and slow cooling in order to remove internal stresses and to make the material less brittle.

Annealing - A process for preventing brittleness in a metal part. The process consists of raising the temperature of the metal to a pre-established level and slowly cooling the steel at a prescribed rate.

Anode - The collector of electrons in an electron tube. Also known as plate; positive electrode.

Anode - The positively charged electrode in an electrochemical process.

Anodizing - The production of a protective oxide film on aluminum or other light metal by passing a high voltage electric current through a bath in which the metal is suspended.

Aquifer - Water bearing stratum.

Ash - The solid residue left after complete combustion.

Assembly - The fitting together of manufactured parts into a complete machine, structure, or unit of a machine.

Assembly - The fitting together of previously manufactured parts or components into a complete machine, unit of a machine, or structure.

Atmospheric Evaporation - Evaporation at ambient pressure utilizing a tower filled with packing material. Air is drawn in from the bottom of the tower and evaporates feed material entering from the top. There is no recovery of the vapors.

Atomic Adsorption - Quantitative chemical instrumentation used for the analysis of elemental constituents.

Automatic Plating - 1. Full - Plating in which the workpieces are automatically conveyed through successive cleaning and plating tanks. 2. Semi - Plating in which the workpieces are conveyed automatically through only one plating tank.

Automatic Wash Water Controls - Automatic solenoid operated shutoff devices which completely stop the flow of water into the processor when it is not being used, thereby avoiding excessive wash water flows.

Autotransformer - A power transformer having one continuous winding that is tapped; part of the winding serves as the primary coil and all of it serves as the secondary coil, or vice versa.

Ballast - A circuit element that serves to limit an electric current or to provide a startling voltage, as in certain types of lamps, such as in fluorescent ceiling fixtures.

Barrel Finishing - The process of polishing a workpiece using a rotating or vibrating container and abrasive grains or other polishing materials to achieve the desired surface appearance.

Basis Metal or Material - That substance of which the workpieces are made and that receives the electroplate and the treatments in preparation for plating.

Batch Treatment - A waste treatment method where wastewater is collected over a period of time and then treated prior to discharge.

Bending - Turning or forcing by a brake press or other device from a straight or even to a curved or angular condition.

Best Available Technology Economically Achievable (BAT) - Level of technology applicable to effluent limitations to be achieved by 1984 for industrial discharges to surface waters as defined by Section 301 (b) (2) (A) of the Act.

Best Practicable Control Technology Currently Available - Level of technology applicable to effluent limitations to be achieved for industrial discharges to surface waters as defined by Section 301 (b) (1) (A) of the Act.

Binder - A material used to promote cohesion between particles of carbon or graphite to produce solid carbon and graphite rods or pieces.

Biochemical Oxygen Demand (BOD) - The amount of oxygen in milligrams per liter used by microorganisms to consume biodegradable organics in wastewater under aerobic conditions.

Biodegradability - The susceptibility of a substance to decomposition by microorganisms; specifically, the rate at which compounds may be chemically broken down by bacteria and/or natural environmental factors.

Biological Wastewater Treatment - Forms of wastewater treatment in which bacteria or biochemical action is intensified to stabilize, oxidize, and nitrify the unstable organic matter present. Intermittent sand filters, contact beds, trickling filters, and activated sludge processes are examples.

Black and White Film - This film consists of a support, usually a plastic film which is coated with a light sensitive emulsion and an outer protective layer. The emulsion contains: gelatin, silver salts of bromide, iodide, chloride, sensitizers, hardeners, and emulsion plasticizers.

Blanking - Cutting desired shapes out of sheet metal by means of dies.

Bleach - A step in color film processing whereby the silver image is converted back to silver halides.

Bleach-Fix or Blix - A solution used in some color processing that functions both as a bleach and as a fix.

Blowdown - The minimum discharge of recirculating water for the purpose of discharging materials contained in the water, the further buildup of which would cause concentration in amounts exceeding limits established by best engineering practice.

BOD5 - The five-day Biochemical Oxygen Demand (BOD5) is the quantity of oxygen used by bacteria in consuming organic matter in a sample of wastewater over a five-day period. BOD from the standard five-day test equals about two-thirds of the total BOD. See Biochemical Oxygen Demand.

Bonding - The process of uniting using an adhesive or fusible ingredient.

Boring - Enlarging a hole by removing metal with a single or occasionally a multiple point cutting tool moving parallel to the axis of rotation of the work or tool. 1. Single-Point Boring - Cutting with a single-point tool. 2. Precision Boring - Cutting to tolerances held within narrow limits. 3. Gun Boring - Cutting of deep holes. 4. Jig Boring - Cutting of high-precision and accurate location holes. 5. Groove Boring - Cutting accurate recesses in hole walls.

Brazing - Joining metals by flowing a thin layer, capillary thickness, of non-ferrous filler metal into the space between them. Bonding results from the intimate contact produced by the dissolution of a small amount of base metal in the molten filler metal, without fusion of the base metal. The term brazing is used where the temperature exceeds 425 C (800 F).

Bright Dipping - The immersion of all or part of a workpiece in a media designed to clean or brighten the surface and leave a protective surface coating on the workpiece.

Brine - An aqueous salt solution.

Buffing - An operation to provide a high luster to a surface. The operation, which is not intended to remove such material, usually follows polishing.

Buffing Compounds - Abrasive contained by a liquid or solid binder composed of fatty acids, grease, or tallow. The binder serves as lubricant, coolant, and an adhesive of the abrasive to the buffing wheel.

Burnishing - Finish sizing and smooth finishing of a workpiece (previously machined or ground) by displacement, rather than removal, of minute surface irregularities with smooth point or line-contact, fixed or rotating tools.

Busbar - A heavy rigid, metallic conductor, usually uninsulated, used to carry a large current or to make a common connection between several circuits.

Bushing - An insulating structure including a central conductor, or providing a central passage for a conductor, with provision for mounting on a barrier (conducting or otherwise), for the purpose of insulating the conductor from the barrier and conducting current from one side of the barrier to the other.

Calcining - To heat to a high temperature without melting or fusing, as to heat unformed ceramic materials in a kiln, or to heat ores, precipitates, concentrates or residues so that hydrates, carbonates or other compounds are decomposed and volatile material is expelled, e.g., to heat limestone to make lime.

Calendering - Process of forming a continuous sheet by squeezing the material between two or more parallel rolls to impart the desired finish or to insure uniform thickness.

Calibration - The determination, checking, or correction of the graduation of any instrument giving quantitative measurements.

Capacitance - The ratio of the charge on one of the plates of a capacitor to the potential difference between the plates.

Capacitor - An electrical circuit element used to store charge temporarily, consisting in general of two conducting materials separated by a dielectric material.

Capital Recovery Costs - Allocates the initial investment and the interest to the total operating cost. The capital recovery cost is equal to the initial investment multiplied by the capital recovery factor.

Captive Operation - A manufacturing operation carried out in a facility to support subsequent manufacturing, fabrication, or assembly operations.

Carbides - Usually refers to the general class of pressed and sintered tungsten carbide cutting tools which contain tungsten carbide plus smaller amounts of titanium and tantalum carbides along with cobalt which acts as a binder. (It is also used to describe hard compounds in steels and cast irons.)

Carbon - A nonmetallic, chiefly tetravalent element found native or as a constituent of coal, petroleum, asphalt, limestone, etc.

Carbon Adsorption - Activated carbon contained in a vessel and installed in either a gas or liquid stream to remove organic contaminants. Carbon is regenerable when subject to steam which forces contaminant to desorb from media.

Carbon Bed Catalytic Destruction - A non-electrolytic process for the catalytic oxidation of cyanide wastes using filters filled with low-temperature coke.

Carcinogen - Substance which causes cancerous growth.

Case Hardening - A heat treating method by which the surface layer of alloys is made substantially harder than the interior. (Carburizing and nitriding are common ways of case hardening steels.)

Cast - A state of the substance after solidification of the molten substance.

Catalytic Bath - A bath containing a substance used to accelerate the rate of chemical reaction.

Category - Also point source category. A segment of industry for which a set of effluent limitations has been established.

Cathode - The negatively charged electrode in an electrochemical process.

Cathode Ray Tube - An electron-beam tube in which the beam can be focused to a small cross section on a luminescent screen and varied in position and intensity to produce a visible pattern.

Cation - The positively charged ions in a solution.

Caustic - Capable of destroying or eating away by chemical action. Applies to strong bases and characterized by the presence of hydroxyl ions in solution.

Caustic Soda - Sodium hydroxide, NaOH, whose solution in water is strongly alkaline.

Cementation - The electrochemical reduction of metal ions by contact with a metal of higher oxidation potential. It is usually used for the simultaneous recovery of copper and reduction of hexavalent chromium with the aid of scrap iron.

Centerless Grinding - Grinding the outside or inside of a workpiece mounted on rollers rather than on centers. The workpiece may be in the form of a cylinder or the frustrum of a cone.

Central Treatment Facility - Treatment plant which co-treats process wastewaters from more than one manufacturing operation or co-treats process wastewaters with non-contact cooling water, or with non-process wastewaters (e.g., utility blowdown, miscellaneous runoff, etc.).

Central Treatment Facility - Treatment plant which co-treats process wastewaters from more than one manufacturing operation or co-treats process wastewaters with noncontact cooling water or with non-process wastewaters (e.g., utility blow-down, miscellaneous runoff, etc.)

Centrifugation - An oil recovery step employing a centrifuge to remove water from waste oil.

Centrifuge - A device having a rotating container in which centrifugal force separates substances of differing densities.

Centrifuge - The removal of water in a sludge and water slurry by introducing the water and sludge slurry into a centrifuge. The sludge is driven outward with the water remaining near the center. The dewatered sludge is usually landfilled.

Ceramic - A product made by the baking or firing of a nonmetallic mineral such as tile, cement, plaster, refractories, and brick.

Chelated Compound - A compound in which the metal is contained as an integral part of a ring structure and is not readily ionized.

Chelating Agent - A coordinate compound in which a central atom (usually a metal) is joined by covalent bonds to two or more other molecules or ions (called ligands) so that heterocyclic rings are formed with the central (metal) atom as part of each ring. Thus, the compound is suspending the metal in solution.

Chemical Brightening - Process utilizing an addition agent that leads to the formation of a bright plate or that improves the brightness of the deposit.

Chemical Coagulation - The destabilization and initial aggregation of colloidal and finely divided suspended matter by the addition of a floc-forming chemical.

Chemical Deposition - Process used to deposit a metal oxide on a substrate. The film is formed by hydrolysis of a mixture of chlorides at the hot surface of the substrate. Careful control of the water mixture insures that the oxide is formed on the substrate surface.

Chemical Etching - To dissolve a part of the surface of a metal or all of the metal laminated to a base.

Chemical Machining - Production of derived shapes and dimensions through selective or overall removal of metal by controlled chemical attack or etching.

Chemical Metal Coloring - The production of desired colors on metal surfaces by appropriate chemical or electrochemical action.

Chemical Milling - Removing large amounts of stock by etching selected areas of complex workpieces. This process entails cleaning, masking, etching, and demasking.

Chemical Oxidation - (Including Cyanide) The addition of chemical agents to wastewater for the purpose of oxidizing pollutant material.

Chemical Oxygen Demand (COD) - The amount of oxygen in milligrams per liter to oxidize both organic and oxidizable inorganic compounds.

Chemical Precipitation - A chemical process in which a chemical in solution reacts with another chemical introduced to that solution to form a third substance which is partially or mainly insoluble and, therefore, appears as a solid.

Chemical Prewash - A salt bath between the fix and final wash which chemically removes the fix from the emulsion at a faster rate than can be done by washing, thereby reducing the after-fix wash water time and volume.

Chemical Recovery Systems - Chemical treatment to remove metal or other materials from wastewater for later reuse.

Chemical Reduction - A chemical reaction in which one or more electrons are transferred to the chemical being reduced from the chemical initiating the transfer (reducing agent).

Chemical Treatment - Treating contaminated water by chemical means.

Chip Dragout - Cutting fluid or oil adhering to metal chips from a machining operation.

Chlorinated Hydrocarbons - Organic compounds containing chlorine such as many insecticides.

Chromate Conversion Coating - Protective coating formed by immersing metal in an aqueous acidified solution consisting substantially of chromic acid or water soluble salts of chromic acid together with various catalysts or activators.

Chromatizing - To treat or impregnate with a chromate (salt of ester of chromic acid) or dichromate, especially with potassium dichromate.

Chrome-Pickle Process - Forming a corrosion-resistant oxide film on the surface of magnesium base metals by immersion in a bath of an alkaline bichromate.

Chromium - A metallic element whose compounds are used in some photographic processes as bleaching or hardening agents.

Circuit Breaker - Device capable of making, carrying, and breaking currents under normal or abnormal circuit conditions.

Clarification - The composite wastewater treatment process consisting of flash mixing of coagulants, pH adjusting chemicals, and/or polyelectrolytes, flocculation, and sedimentation.

Clarifier - A unit which provides for settling and removal of solids from wastewater.

Cleaning - The removal of soil and dirt (including grit and grease) from a workpiece using water with or without a detergent or other dispersing agent.

Clearing Bath - A processing solution that removes most residual fixer from processed film or paper prior to washing, minimizing the water requirement.

Chlorination - The application of chlorine to water generally for purposes of disinfection, but frequently for accomplishing other biological or chemical results.

Closed-Loop Evaporation System - A system used for the recovery of chemicals and water from a chemical finishing process. An evaporator concentrates flow from the rinse water holding tank. The concentrated rinse solution is returned to the bath, and distilled water is returned to the final rinse tank. The system is designed for recovering 100 percent of chemicals normally lost in dragout for reuse in the process.

Closed-Loop Rinsing - The recirculation of rinse water without the introduction of additional makeup water.

Coagulation - A chemical reaction in which polyvalent ions neutralize the repulsive charges surrounding colloidal particles.

COD - See Chemical Oxygen Demand

Coil - A number of turns of wire used to introduce inductance into an electric circuit, to produce magnetic flux, or to react mechanically to a changing magnetic flux.

Coil-Core Assembly - A unit made up of the coil windings of a transformer placed over the magnetic core.

Cold Drawing - A process of forcing material through dies or other mandrels to produce wire, rod, tubular and some bars.

Cold Heading - A method of forcing metal to flow cold into enlarged sections by endwise squeezing. Typical coldheaded parts are standard screws, bolts under 1 in. diameter and a large variety of machine parts such as small gears with stems.

Cold Rolling - A process of forcing material through rollers to produce bars and sheet stock.

Colloids - A finely divided dispersion of one material called the "dispersed phase" (solid) in another material called the "dispersion medium" (liquid). Normally negatively charged.

Colorimetric - A procedure for establishing the concentration of impurities in water by comparing its color to a set of known color impurity standards.

Color Couplers - A group of organic chemicals which react with the oxidized components of the developers to form color dyes. They are either incorporated in the film emulsion at the time of manufacture or they are included in the color developing solution.

Color Film - Color film has three separate light sensitive emulsion layers, which after inclusion of the appropriate sensitizing dyes, record an image of the blue light components on one layer, the green light components on another, and the red light components on the third layer.

Color Reversal (DC) Process - A color reversal film process in which the color couplers are added during development.

Color Reversal (IC) Process - A color reversal film and paper process in which the color couplers which form the color dye images are incorporated into the emulsion layers at the time of manufacture.

Common Metals - Copper, nickel, chromium, zinc, tin, lead, cadmium, iron, aluminum, or any combination thereof.

Compatible Pollutants - Those pollutants which can be adequately treated in publicly-owned treatment works without upsetting the treatment process.

Complex Cyanide - This term refers to a complex ion containing cyanide ions and a cation such as iron, e.g., ferrocyanide and/or ferricyanide.

Complexing Agent - A compound that will join with a metal to form an ion which has a molecular structure consisting of a central atom (the metal) bonded to other atoms by coordinate covalent bonds.

Composite Wastewater Sample - A combination of individual samples of water or wastewater taken at selected intervals and mixed in proportion to flow or time to minimize the effect of the variability of an individual sample.

Composite Wastewater Sample - A combination of individual samples of water or wastewater taken at selected intervals, generally hourly for some specified period, to minimize the effect of the variability of the individual sample. Individual samples may have equal volume or may be proportioned to the flow at time of sampling.

Concentric Windings - Transformer windings in which the low-voltage winding is in the form of a cylinder next to the core, and the high-voltage winding, also cylindrical, surrounds the low-voltage winding.

Conductance - See Electrical Conductivity.

Conductivity Meter - An instrument which displays a quantitative indication of conductance.

Conductivity Surface - A surface that can transfer heat or electricity.

Conductor - A wire, cable, or other body or medium suitable for carrying electric current.

Conduit - Tubing of flexible metal or other material through which insulated electric wires are run.

Conservation - Methods applied to make maximum use of processing chemicals and water and to keep the quantity of effluent discharged to a minimum.

Contact Water - See Process Wastewater.

Contamination - A general term signifying the introduction into water of microorganisms, chemicals, wastes or sewage which renders the water unfit for its intended use.

Continuous Length Processor - An automatic processing machine whereby long rolls of film or paper are fed into successive photoprocessing tanks via a series of appropriate crossover connections between racks. The starting end of the material to be processed is attached to a leader which guides the material through the machine.

Continuous Treatment - Treatment of waste streams operating without interruption as opposed to batch treatment; sometimes referred to as flow-through treatment.

Contractor Removal - Disposal of oils, spent solutions, or sludge by a scavenger service.

Conventional Silver Recovery - The use of metallic replacement of electrolytic methods or both for the recovery of silver from fix solutions.

Conversion Coating - A coating produced by chemical or electrochemical treatment of a metallic surface that gives a superficial layer containing a compound of the metal. For example, chromate coating on zinc and cadmium, oxide coatings on steel.

Coolant - See Cutting Fluids.

Cooling Tower - A device used to cool manufacturing process water before returning the water for reuse.

Cooling Water - Water which is used to adsorb and transport heat generated in a process or machinery.

Copper - A common, reddish, chiefly univalent and bivalent metallic element that is ductile and malleable and one of the best conductors of heat and electricity.

Copper Flash - Quick preliminary deposition of copper for making surface acceptable for subsequent plating.

Coprecipitation of Metals - Precipitation of a metal with another metal.

Core (Magnetic Core) - A quantity of ferrous material placed in a coil or transformer to provide a better path than air for magnetic flux, thereby increasing the inductance of the coil or increasing the coupling between the windings of a transformer.

Corona Discharge - A discharge of electricity appearing as a bluish-purple glow on the surface of and adjacent to a conductor when the voltage gradient exceeds a certain critical value; caused by ionization of the surrounding air by the high voltage.

Corrosion Resistant Steels - A term often used to describe the stainless steels with high nickel and chromium alloy content.

Cost of Capital - Capital recovery costs minus the depreciation.

Counterboring - Removal of material to enlarge a hole for part of its depth with a rotary, pilot guided, end cutting tool having two or more cutting lips and usually having straight or helical flutes for the passage of chips and the admission of a cutting fluid.

Countercurrent Rinsing - Rinsing of parts in such a manner that the rinse water is removed from tank to tank counter to the flow of parts being rinsed.

Countercurrent Washing - A method of washing film or paper using a segmented tank system in which water is cascaded progressively from one tank segment to the next counter to the movement of the film or paper.

Countersinking - Beveling or tapering the work material around the periphery of a hole creating a concentric surface at an angle less than 90 degrees with the centerline of the hole for the purpose of chamfering holes or recessing screw and rivet heads.

Crystalline Solid - A substance with an ordered structure, such as a crystal.

Crystallization - 1. Process used to manufacture semiconductors in the electronics industry. 2. A means of concentrating pollutants in wastewaters by crystallizing out of pure water.

Curing - A heating/drying process carried out in an elevated-temperature enclosure.

Current Carrying Capacity - The maximum current that can be continuously carried without causing permanent deterioration of electrical or mechanical properties of a device or conductor.

Cutting Fluids - Lubricants employed to ease metal and machining operations, produce surface smoothness and extend tool life by providing lubricity and cooling. Fluids can be emulsified oils in water, straight mineral oils when better smoothness and accuracy are required, or blends of both.

Cyaniding - A process of case hardening an iron-base alloy by the simultaneous adsorption of carbon and nitrogen by heating in a cyanide salt. Cyaniding is usually followed by quenching to produce a hard case.

Cyclone Separator - A device which removes entrained solids from gas streams.

Dag (Aquadag) - A conductive graphite coating on the inner and outer side walls of some cathode ray tubes.

Dead Rinse - A rinse step in which water is not replenished or discharged.

Deburring - Removal of burrs or sharp edges from parts by filing, grinding or rolling the work in a barrel with abrasives suspended in a suitable medium.

Deep Bed Filtration - The common removal of suspended solids from wastewater streams by filtering through relatively deep (o.3-0.9 m) granular bed. The porous bed formed by the granular media can be designed to remove practically all suspended particles by physical-chemical effects.

Degassing - (Fluxing) The removal of hydrogen and other impurities from molten primary aluminum in a casthouse holding furnace by injecting chlorine gas (often with nitrogen and carbon).

Degradable - That which can be reduced, broken down or chemically separated.

Degreasing - The process of removing grease and oil from the surface of the basis material.

Demineralization - The removal from water of mineral contaminants usually present in ionized form. The methods used include ion-exchange techniques, flash distillation or electrolysis.

Denitrification (Biological) - The reduction of nitrates to nitrogen gas by bacteria.

Deoxidizing - The removal of an oxide film from an alloy such as aluminum oxide.

Depreciation - Decline in value of a capital asset caused either by use or by obsolescence.

Descaling - The removal of scale and metallic oxides from the surface of a metal by mechanical or chemical means. The former includes the use of steam, scale-breakers and chipping tools, the latter method includes pickling in acid solutions.

Desmutting - The removal of smut (matter that soils or blackens) generally by chemical action.

Develop - A step in photoprocessing whereby the latent image is made visible in a developer solution.

Developer - A chemical processing solution containing a developing agent. This solution converts the exposed portions of the photographic emulsion to silver, creating images of metallic silver.

Developing Agents - These photographic materials usually are aromatic compounds with phenolic or amino electron-donor groups arranged ortho or para with respect to each other, such as: hydroquinone, methyl p-amino-phenol (metol) or 1-phenol-3 pyrazolidone (phenidone).

Dewatering - A process in which water is removed from sludge.

Di-n-octyl-phthalate - A liquid dielectric that is presently being substituted for a PCB dielectric fluid.

Diaminobenzidene - A chemical used in the standard method of measuring the concentrations of selenium in a solution.

Dibasic Acid - An acid capable of donating two protons (hydrogen ions).

Dichromate Bleach - A bleach used in some black and white reversal and color film processing.

Dichromate Reflux - A standard method of measuring the chemical oxygen demand of a solution.

Dicing - Sawing or otherwise machining a semiconductor wafer into small squares or dice from which transistors and diodes can be fabricated.

Die - A tool or mold used to cut shapes to or form impressions on materials such as metals and ceramics.

Die Casting - (hot chamber, vacuum, pressure) Castings are produced by forcing molten metal under pressure into metal molds called dies. in hot chamber machines, the pressure cylinder is submerged in the molten metal resulting in a minimum of time and metal cooling during casting. Vacuum feed machines use a vacuum to draw a measured amount of melt from the molten bath into the feed chamber. Pressure feed systems use a hydraulic or pneumatic cylinder to feed molten metal to the die.

Die Cutting (Also Blanking) - Cutting of plastic or metal sheets into shapes by striking with a punch.

Dielectric - A material that is highly resistant to the conductance of electricity; an insulator.

Digestion - A standard method of measuring organic nitrogen.

Diode (Semiconductor), (Also Crystal Diode, Crystal Rectifier) - A two-electrode semiconductor device that utilizes the rectifying properties of a p-n junction or point contact.

Dip and Dunk - An automatic processing machine whereby strips of film are "dipped" into successive photoprocessing tanks and held for the appropriate time.

Dipping - Material coating by briefly immersing parts in a molten bath, solution or suspension.

Direct Discharge - The discharge of wastewater to waters of the United States.

Direct Labor Costs - Salaries, wages and other direct compensations earned by the employee.

Discharge of pollutant(s) - 1. The addition of any pollutant to navigable waters from any point source. 2. Any addition of any pollutant to the waters of the contiguous zone or the ocean from any point source, other than from a vessel or other floating craft. The term "discharge" includes either the discharge of a single pollutant or the discharge of multiple pollutants.

Discrete Device - Individually manufactured transistor, diode, etc.

Dispersed-air Flotation - Separation of low density contaminants from water using minute air bubbles attached to individual particles to provide or increase the buoyancy of the particle. The bubbles are generated by introducing air through a revolving impeller or porous media.

Dissolved Oxygen (DO) - The oxygen dissolved in sewage, water, or other liquid, usually expressed in milligrams per liter or percent of saturation. It is the test used in BOD determination.

Dissolved-air Flotation - Separation of low density contaminants from water using minute air bubbles attached to individual particles to provide or increase the buoyancy of the particle. The air is put into solution under elevated pressure and later released under atmospheric pressure or put into solution by aeration at atmospheric pressure and then released under a vacuum.

Dissolved Solids - Solid matter in effluent that will not settle out or separate with filtration.

Distillation - Vaporization of a liquid followed by condensation of the vapor.

Distillation Refining - A metal with an impurity having a higher vapor pressure than the base metal can be refined by heating the metal to the point where the impurity vaporizes.

Distillation-SPADNS - A standard method of measuring the concentration of fluoride in a solution.

Distribution Transformer - An element of an electric distribution system located near consumers which changes primary distribution voltage to a lower consumer voltage.

Dopant - An impurity element added to semiconductor materials used in crystal diodes and transistors.

Drag-In - Water or solution carried into another solution by the work and the associated handling equipment.

Dragout - The solution that adheres to the part or workpiece and is carried past the edge of the tank.

Dragout Reduction - Minimization of the amount of material (bath or solution) removed from a process tank by adhering to the part or its transfer device.

Drainage Phase - Period in which the excess plating solution adhering to the part or workpiece is allowed to drain off.

Drawing - Reduction of cross section area and increasing the length by pulling metal through conical taper dies.

Drilling - Hole making with a rotary, end-cutting tool having one or more cutting lips and one or more helical or straight flutes or tubes for the ejection of chips and the passage of a cutting fluid. 1. Center Drilling - Drilling a conical hole in the end of a workpiece. 2. Core Drilling - Enlarging a hole with a chamfer-edged, multiple-flute drill. 3. Spade Drilling - Drilling with a flat blade drill tip. 4. Step Drilling - Using a multiple diameter drill. 5. Gun Drilling - Using special straight flute drills with a single lip and cutting fluid at high pressures for deep hole drilling. 6. Oil Hole or Pressurized Coolant Drilling - Using a drill with one or more continuous holes through its body and shank to permit the passage of a high pressure cutting fluid which emerges at the drill point and ejects chips.

Drip Station - Empty tank over which parts are allowed to drain freely to decrease end dragout.

Drip Time - The period during which a part is suspended over baths in order to allow the excessive dragout to drain off.

Dry - The final processing step which involves drying the photographic film or paper in a dust-free atmosphere.

Dry Electrolytic Capacitor - An electrolytic capacitor with a paste rather than liquid electrolyte.

Dry Slug - Usually refers to a plastic-encased sintered tantalum slug type capacitor.

Dry Transformer - Having the core and coils neither impregnated with an insulating fluid nor immersed in an insulating oil.

Drying Beds - Areas for dewatering of sludge by evaporation and seepage.

Dye Image - A color image formed when the oxidized developer combines with the color couplers.

EDTA Titration - EDTA - ethylenediamine tetraacetic acid (or its salts). A standard method of measuring the hardness of a solution.

Effluent - The water and the quantities, rates, and concentrations of chemical, physical, biological, and other constituents which are discharged from point sources.

Effluent Limitation - Any restriction (including schedules of compliance) established by a state or the federal EPA on quantities, rates, and concentrations of chemical, physical, biological, and other constituents which are discharged from point sources into navigable waters, the waters of the contiguous zone, or the ocean.

Electrical Conductivity - The property which allows an electric current to flow when a potential difference is applied. It is the reciprocal of the resistance in ohms measured between opposite faces of a centimeter cube of an aqueous solution at a specified temperature. It is expressed as microohms per centimeter at temperature degrees Celsius.

Electrobrightening - A process of reversed electro-deposition which results in anodic metal taking a high polish.

Electrochemical Machining - Shaping of an anode by the following process: The anode and cathode are placed close together and electrolyte is pumped into the space between them. An electrical potential is applied to the electrodes causing anode metal to be dissolved selectively, producing a shaped anode that complements the shape of the cathode.

Electrocleaning - The process of anodic removal of surface oxides and scale from a workpiece.

Electrode - Conduction material for passing electric current into or out of a solution by adding electrons to or taking electrons from ions in the solution.

Electrodialysis - A treatment process that uses electrical current and an arrangement of permeable membranes to separate soluble minerals from water. Often used to desalinate salt or brackish water.

Electroless Plating - Deposition of a metallic coating by a controlled chemical reduction that is catalyzed by the metal or alloy being deposited.

Electrolysis - The chemical decomposition by an electric current of a substance in a dissolved or molten state.

Electrolyte - A liquid, most often a solution, that will conduct an electric current.

Electrolytic Cell - A unit apparatus in which electromechanical reactions are produced by applying electrical energy or which supplies electrical energy as a result of chemical reactions and which includes two or more electrodes and one or more electrolytes contained in a suitable vessel.

Electrolytic Decomposition - An electrochemical treatment used for the oxidation of cyanides. The method is practical and economical when applied to concentrated solutions such as contaminated baths, cyanide dips, stripping solutions, and concentrated rinses. Electrolysis is carried out at a current density of 35 amp/sq. ft. at the anode and 70 amp/sq. ft. at the cathode. Metal is deposited at the cathode and can be reclaimed.

Electrolytic Oxidation - A reaction by an electrolyte in which there is an increase in valence resulting from a loss of electrons.

Electrolytic Reduction - A reaction in which there is a decrease in valence resulting from a gain in electrons.

Electrolytic Refining - The method of producing pure metals by making the impure metal the anode in an electrolytic cell and depositing a pure cathode. The impurities either remain undissolved at the anode or pass into solutions in the electrolyte.

Electromechanical Machining (ECM) - A machining process whereby the part to be machined is made the anode and a shaped cathode is maintained in close proximity to the work. Electrolyte is pumped between the electrodes and a potential applied with the result that metal is rapidly dissolved from the workpiece in a selective manner and the shape produced on the workpiece complements that of the cathode.

Electrolytic Silver Recovery - The removal of silver from silver-bearing solutions by application of a direct current to electrodes in the solution causing metallic silver to deposit on the cathode.

Electrometallurgical Process - The application of electric current to a metallurgical process either for electrolytic deposition or as a source of heat.

Electrometric Titration - A standard method of measuring the alkalinity of a solution.

Electron Beam Lithography - Similar to photolithography - A fine beam of electrons is used to scan a pattern and expose an electron-sensitive resist in the unmasked areas of the object surface.

Electron Beam Machining - The process of removing material from a workpiece by a high velocity focused stream of electrons which melt and vaporize the workpiece at the point of impingerence.

Electron Discharge Lamp - An electron lamp in which light is produced by passage of an electric current through a metallic vapor or gas.

Electron Gun - An electrode structure that produces and may control, focus, deflect and converge one or more electron beams in an electron tube.

Electron Tube - An electron device in which conduction of electricity is accomplished by electrons moving through a vacuum or gaseous medium within a gas-tight envelope.

Electroplating - The production of a thin coating of one metal on a surface by electrodeposition.

Electropolishing - Electrolytic corrosion process that increases the percentage of specular reflectance from a metallic surface.

Elution - 1. The process of washing out, or removing with the use of a solvent. 2. In an ion exchange process it is defined as the stripping of adsorbed ions from an ion exchange resin by passing through the resin solutions containing other ions in relatively high concentrations.

Embossing - Raising a design in relief against a surface.

Emissive Coating - An oxide coating applied to an electrode to enhance the emission of electrons.

Emulsified Oil and Grease - An oil or grease dispersed in an immiscible liquid usually in droplets of larger than colloidal size. In general suspension of oil or grease within another liquid (usually water).

Emulsifying Agent - A material that increases the stability of a dispersion of one liquid in another.

Emulsion Breaking - Decreasing the stability of dispersion of one liquid in another.

Emulsion Cleaning - A cleaning process using organic solvents dispersed in an aqueous medium with the aid of an emulsifying agent.

End-of-Pipe Treatment - The reduction and/or removal of pollutants by treatment just prior to actual discharge.

Environmental Protection Agency (EPA) - The United States Environmental Protection Agency.

Epitaxial Layer - A (thin) semiconductor layer having the same crystalline orientation as the substrate on which it is grown.

Epitaxial Transistor - Transistor with one or more epitaxial layers.

Equalization - (Continuous Flow) - The balancing of flow or pollutant load using a holding tank for a system that has widely varying inflow rates.

Equilibrium Concentration - A state at which the concentration of chemicals in a solution remain in a constant proportion to one another.

Etchant - The material used in the chemical process of removing glass fibers and epoxy between neighboring conductor layers of a PC board for a given distance.

Etching - A process where material is removed by chemical action.

Evaporation - A technique used to concentrate solids by removing water resulting in a quantity of concentrated waste containing the solids.

Evaporation Ponds - Liquid waste disposal areas that allow the liquid to vaporize to cool discharge water temperatures or to thicken sludge.

Extrusion - A material that is forced through a die to form lengths of rod, tube or special sections.

Fermentation - A chemical change to break down biodegradable waste. The change is induced by a living organism or enzyme, specifically bacteria or microorganisms occurring in unicellular plants such as yeast, molds, or fungi.

Ferricyanide - This ion, usually in the form of potassium ferricyanide, is used as a bleach for oxidizing metallic silver to ionic silver in some color processes. Ferricyanide is reduced to ferrocyanide as it oxidized silver in the film emulsions.

Ferricyanide Bleach - A processing solution containing the ferricyanide ion. This is used to convert metallic silver to ionic silver, which is removed in the fixing step.

Ferrocyanide - The ion that results when ferricyanide oxidizes silver or reacts with various reducing agents.

Field-Effect Transistors - Transistors made by the metal-oxide-semiconductor (MOS) technique, differing from bipolar ones in that only one kind of charge carrier is active in a single device. Those that employ holes are p-MOS transistors.

Filament - 1. Metallic wire which is heated in an incandescent lamp to produce light by passing an electron current through it. 2. A cathode in a fluorescent lamp that emits electrons when electric current is passed through it.

Filtering Capacitor - A capacitor used in a power-supply filter system to provide a low-reactance path for alternating currents and thereby suppress ripple currents, without affecting direct currents.

Filtrate - Liquid after passing through a filter.

Filtration - Removal of solid particles from liquid or particles from air or gas stream by means of a permeable membrane. Types: Gravity, Pressure, Microstraining, Ultrafiltration, Reverse Osmosis (Hyperfiltration).

Fix - A step in photoprocessing whereby the silver halides are removed from the emulsion using a solvent such as sodium thiosulfate.

Fixed Capacitor - A capacitor having a definite capacitance value that cannot be adjusted.

Flame Hardened - Surface hardened by controlled torch heating followed by quenching with water or air.

Flame Spraying - The process of applying a metallic coating to a workpiece whereby finely powered powered fragments or wire, together with suitable fluxes, are projected through a cone of flame onto the workpiece.

Flameless Atomic Absorption - A method of measuring low concentration values of certain metals in a solution.

Flash Evaporation - Evaporation using steam heated tubes with feed material under high vacuum. Feed material "flashes off" when it enters the evaporation chamber.

Float Gauge - A device for measuring the elevation of the surface of a liquid, the actuating element of which is a buoyant float that rests on the surface of the liquid and rises or falls with it. The elevation of the surface is measured by a chain or tape attached to the float.

Floc - A very fine, fluffy mass formed by the aggregation of fine suspended particles.

Flocculation - The process of separating suspended solids from wastewater by chemical creation of clumps of flocs.

Flocculator - An apparatus designed for the formation of floc in water or sewage.

Flotation - The process of removing finely divided particles from a liquid suspension by attaching gas bubbles to the particles, increasing their buoyancy, and thus concentrating them at the surface of the liquid medium.

Flow-proportioned Sample - A sampled stream whose pollutants are apportioned to contributing streams in proportion to the flow rates of the contributing streams.

Fluorescent Lamp - An electric discharge lamp in which phosphor materials transform ultraviolet radiation from mercury vapor ionization to visible light.

Fluxing - (Degassing) The removal of oxides and other impurities from molten primary aluminum in a casthouse holding furnace by injecting chlorine gas (often with nitrogen and carbon monoxide).

Fog - A type of rinse consisting of a fine spray.

Forming - Application of voltage to an electrolytic capacitor, electrolytic rectifier or semiconductor device to produce a desired permanent change in electrical characteristics as part of the manufacturing process.

Forming Compounds (Sheet) - Tightly adhering lubricants composed of fatty oils, fatty acids, soaps, and waxes and designed to resist the high surface temperatures and pressures the metal would otherwise experience in forming.

Forming Compounds (Wire) - Tightly adhering lubricants composed of solids (white lead, talc, graphite, or molybdenum disulfide) and soluble oils for cooling and corrosion protection. Lubricants typically contain sulfer, chlorine, or phosphate additives.

Free Cyanide - 1. True - the actual concentration of cyanide radical or equivalent alkali cyanide not combined in complex ions with metals in solutions. 2. Calculated - the concentration of cyanide or alkali cyanide present in solution in excess of that calculated as necessary to form a specified complex ion with a metal or metals present in solution. 3. Analytical - the free cyanide content of a solution as determined by a specified analytical method.

Freezing/Crystallization - The solidification of a liquid into aggregations of regular geometric forms (crystals) accomplished by subtraction of heat from the liquid. This process can be used for removal of solids, oils, greases, and heavy metals from industrial wastewater.

Frit Seal - A seal made by fusing together metallic powders with a glass binder for such applications as hermetically sealing ceramic packages for integrated circuits.

Funnel - The rear, funnel-shaped portion of the glass enclosure of a cathode ray tube.

Fuse - Overcurrent protective device with a circuit-opening fusible part that would be heated and severed by overcurrent passage.

Galvanizing - The deposition of zinc on the surface of steel for corrosion protection.

Gas Carburizing - The introduction of carbon into the surface layers of mill steel by heating in a current of gas high in carbon.

Gas Chromotography - Chemical analytical instrumentation generally used for quantitative organic analysis.

Gas Nitriding - Case hardening metal by heating and diffusing nitrogen gas into the surface.

Gas Phase Separation - The process of separating volatile constituents from water by the application of selective gas permeable membranes.

Gate - One of the electrodes in a field effect transistor.

Gear Forming - Process for making small gears by rolling the gear material as it is pressed between hardened gear shaped dies.

Getter - A metal coating inside a lamp which is activated by an electric current to absorb residual water vapor and oxygen.

Glass - A hard, amorphous, inorganic, usually transparent, brittle substance made by fusing silicates, and sometimes borates and phosphates, with certain basic oxides and then rapidly cooling to prevent crystallization.

Glass Fiber Filtration - A standard method of measuring total suspended solids.

Glow Lamp - An electronic device, containing at least two electrodes and an inert gas, in which light is produced by a cloud of electrons close to the negative electrode when a voltage is applied between the electrodes.

Good Housekeeping - (In-Plant Technology) Good and proper maintenance minimizing spills and upsets.

GPD - Gallons per day.

Grab Sample - A single sample of wastewater taken without regard to time or flow.

Graphite - A soft black lustrous carbon that conducts electricity and is a constituent of coal, petroleum, asphalt, limestone, etc.

Gravimetric 103-105C - A standard method of measuring total solids in aqueous solutions.

Gravity Filtration - Settling of heavier and rising of lighter constituents within a solution.

Gravity Flotation - The separation of water and low density contaminants such as oil or grease by reduction of the wastewater flow velocity and turbulence for a sufficient time to permit separation due to difference in specific gravity. The floated material is removed by some skimming technique.

Gray Cast Irons - Alloys primarily of iron, carbon and silicon along with other alloying elements in which the graphite is in flake form. (These irons are characterized by low ductility but have many other properties such as good castability and good damping capacity.)

Grease - In wastewater, a group of substances including fats, waxes, free fatty acids, calcium and magnesium soaps, mineral oils, and certain other nonfatty materials. The type of solvent and method used for extraction should be stated for qualification.

Grease Skimmer - A device for removing floating grease or scum from the surface of wastewater in a tank.

Grid - An electrode located between the cathode and anode of an electron tube, which has one or more openings through which electrons or ions can pass, and which controls the flow of electrons from cathode to anode.

Grinding - The process of removing stock from a workpiece by the use of abrasive grains held by a rigid or semi-rigid binder.

Grinding Fluids - Water based, straight oil, or synthetic based lubricants containing mineral oils, soaps, or fatty materials which serve to cool the part and maintain the abrasiveness of the grinding wheel face.

Hammer Forging - Heating and pounding metal to shape it into the desired form.

Harden - This photoprocessing step serves to check emulsion swelling and raise the melting point for the emulsion to allow for drying at a higher temperature.

Hardened - Designates condition produced by various heat treatments such as quench hardening, age hardening and precipitation hardening.

Hardener - A chemical present in some photographic solutions that reacts with gelatin in the emulsion to protect the film from damage during or after processing. Common hardeners are potassium aluminum sulfate, potassium chromium sulfate, and formaldehyde solution.

Hardness - A characteristic of water, imparted by salts of calcium, magnesium and iron such as bicarbonates, carbonates, sulfates, chlorides and nitrates, that cause curdling of soap, deposition of scale, damage in some industrial processes and sometimes objectionable taste. It may be determined by a standard laboratory procedure or computed from the amounts of calcium and magnesium as well as iron, aluminum, manganese, barium, strontium, and zinc and is expressed as equivalent calcium carbonate.

Heat Treatment - The modification of the physical properties of a workpiece through the application of controlled heating and cooling cycles. Such operations are heat treating, tempering, carburizing, cyaniding, nitriding, annealing, normalizing, austenizing, quenching, austempering, siliconizing, martempering, and malleabilizing are included in this definition.

Heavy Metals - Metals which can be precipitated by hydrogen sulfide in acid solution, e.g., lead, silver, gold, mercury, bismuth, copper, nickel, iron, chromium, zinc, cadmium, and tin.

Holding Tank - A reservoir to contain preparation materials so as to be ready for immediate service.

Honing - A finishing operation using fine grit abrasive stones to produce accurate dimensions and excellent finish.

Hot Dip Coating - The process of coating a metallic workpiece with another metal by immersion in a molten bath to provide a protective film.

Hot Rolled - A term used to describe alloys which are rolled at temperatures above the recrystallization temperature. (Many alloys are hot rolled, and machinability of such alloys may vary because of differences in cooling conditions from lot to lot.)

Hot Stamping - Engraving operation for marking plastics in which roll leaf is stamped with heated metal dies onto the face of the plastics. Ink compounds can also be used.

Hybrid Integrated Circuits - A circuit that is part integrated and part discrete.

Hydrofluoric Acid - Hydrogen fluoride in aqueous solution.

Hydrophilic - A surface having a strong affinity for water or being readily wettable.

Hydrophobic - A surface which is non-wettable or not readily wettable.

Hydrostatic Pressure - The force per unit area measured in terms of the height of a column of water under the influence of gravity.

Immersed Area - Total area wetted by the solution or plated area plus masked area.l

Immersion Plate - A metallic deposit produced by a displacement reaction in which one metal displaces another from solution, for example: Fe + Cu (+2) = Cu + Fe (+2).

Impact Deformation - The process of applying impact force to a workpiece such that the workpiece is permanently deformed or shaped. Impact deformation operations such as shot peening, peening, forging, high energy forming, heading, or stamping.

Impact Extrusion - A cold extrusion process for producing tubular components by striking a slug of the metal, which has been placed in the cavity of the die, with a punch moving at high velocity.

Impregnate - To force a liquid substance into the spaces of a porous solid in order to change its properties.

In-Process Control Technology - The regulation and the conservation of chemicals and the reduction of water usage throughout the operations as opposed to end-of-pipe treatment.

Incandescent Lamp - An electric lamp producing light in which a metallic filament is heated white-hot in a vacuum by passage of an electric current through it.

Incineration - (Sludge Disposal) The combustion (by burning) of organic matter in wastewater sludge after dewatering by evaporation.

Incompatible Pollutants - Those pollutants which would cause harm to, adversely affect the performance of, or be inadequately treated in publicly-owned treatment works.

Independent Operation - Job shop or contract shop in which electroplating is done on workpieces owned by the customer.

Industrial User - Any industry that introduces pollutants into public sewer systems and whose wastes are treated by a publicly-owned treatment facility.

Industrial Wastes - The liquid wastes from industrial processes, as distinct from domestic or sanitary wastes.

Influent - Water or other liquid, either raw or partly treated, flowing into a reservoir basin or treatment plant.

Inhibition - The slowing down or stoppage of chemical or biological reactions by certain compounds or ions.

Inspection - A checking or testing of something against standards or specification.

Insulating Paper - A standard material for insulating electrical equipment, usually consisting of bond or kraft paper coated with black or yellow insulating varnish on both sides.

Insulation (Electrical Insulation) - A material having high electrical resistivity and therefore suitable for separating adjacent conductors in an electric circuit or preventing possible future contact between conductors.

Insulator - A nonconducting support for an electric conductor.

Intake Water - Gross water minus reuse water.

Integrated Chemical Treatment - A waste treatment method in which a chemical rinse tank is inserted in the plating line between the process tank and the water rinse tank. The chemical rinse solution is continuously circulated through the tank and removes the dragout while reacting chemicals with it.

Integrated Circuit (IC) - 1. A combination of interconnected circuit elements inseparably associated on or within a continuous substrate. 2. Any electronic device in which both active and passive elements are contained in a single package. Methods of making an integrated circuit are by masking process, screening and chemical deposition.

Intermittent Filter - A natural or artificial bed of sand or other fine-grained material onto which sewage is intermittently flooded and through which it passes, with time allowed for filtration and the maintenance of aerobic conditions.

Ion - An atom or group of atoms possessing an electrical charge.

Ion Exchange - A reversible chemical reaction between a solid (ion exchanger) and a fluid (usually a water solution) by means of which ions may be interchanged from one substance to another. The superficial physical structure of the solid is not affected.

Ion Exchange Resins - Synthetic resins containing active groups (usually sulfonic, carboxylic, phenol, or substituted amino groups) that give the resin the property of combining with or exchanging ions between the resin and a solution.

Ion Implantation - A process of introducing impurities into the near surface regions of solids by directing a beam of ions at the solid.

Iridite Dip Process - Dipping process for zinc or zinc-coated objects that deposits protective film that is a chromium gel, chromium oxide, or hydrated chromium oxide.

Junction - A region of transition between two different semiconducting regions in a semiconductor device such as a p-n junction, or between a metal and a semiconductor.

Junction Box - A protective enclosure into which wires or cables are led and connected to form joints.

Kiln - (Rotary) A large cylindrical mechanized type of furnace.

Klystron - An evacuated electron-beam tube in which an initial velocity modulation imparted to electrons in the beam results subsequently in density modulation of the beam; used as an amplifier in the microwave region or as an oscillator.

Knife Switch - Form of switch where a moving blade enters stationary contact clips.

Knurling - Impressing a design into a metallic surface, usually by means of small, hard rollers that carry the corresponding design on their surfaces.

Lagoon - A man-made pond or lake for holding wastewater for the removal of suspended solids. Lagoons are also used as retention ponds, after chemical clarification to polish the effluent and to safeguard against upsets in the clarifier; for stabilization of organic matter by biological oxidation; for storage of sludge; and for cooling of water.

Laminate - 1. A composite metal, wood or plastic usually in the form of sheet or bar, composed of two or more layers so bonded that the composite forms a structural member. 2. To form a product of two or more bonded layers.

Landfill - Disposal of inert, insoluble waste solids by dumping at an approved site and covering with earth.

Lapping - An abrading process to improve surface quality by reducing roughness, waviness and defects to produce accurate as well as smooth surfaces.

Leach Field - An area of ground to which wastewater is discharged. Not considered an acceptable treatment method for industrial wastes.

Leaching - Dissolving out by the action of a percolating liquid, such as water, seeping through a landfill.

Ligands - The molecules attached to the central atom by coordinate covalent bonds.

Lime - Any of a family of chemicals consisting essentially of calcium hydroxide made from limestone (calcite) which is composed almost wholly of calcium carbonates or a mixture of calcium and magnesium carbonates.

Limiting Orifice - A device that limits flow by constriction to a relatively small area. A constant flow can be obtained over a wide range of upstream pressures.

Low Flow Prewash - A system which concentrates most of the fix carryout in a low volume after-fix prewash tank. The system consists of segmenting the after-fix prewash tank to provide a small prewash section with separate wash water make-up and overflow.

Machining - The process of removing stock from a workpiece by forcing a cutting tool through the workpiece removing a chip of basis material. Machining operations such as turning, milling, drilling, boring, tapping, planing, broaching, sawing and filing, and chamfering are included in this definition.

Maintenance - The upkeep of property or equipment.

Make-up Water - Total amount of water used by any process/process step.

Mandrel - A metal support serving as a core around which the metals are wound and annealed to form a central hole.

Mask (Shadow Mask) - Thin sheet steel screen with thousands of apertures through which electron beams pass to a color picture tube screen. The color of an image depends on the balance from each of three different electron beams passing through the mask.

Masking - The application of a substance to a surface for the prevention of plating to said area.

Mechanical Plating - Providing a coating wherein fine metal powders are peened onto the part by tumbling or other means.

Membrane - A thin sheet of synthetic polymer through the apertures of which small molecules can pass, while larger ones are retained.

Membrane Filtration - Filtration at pressures ranging from 50 to 100 psig with the use of membranes or thin films. The membranes have accurately controlled pore sites and typically low flux rates.

Metallic Replacement - This occurs when a metal such as iron comes in contact with a solution containing dissolved ions of a less active metal such as silver. The dissolved silver ions react with solid metal (iron). The more active metal (iron) goes into solution as an ion and ions of the less active metal become solid metal (silver).

Metal Ion - An atom or radical that has lost or gained one or more electrons and has thus acquired an electric charge. Positively charged ions are cations, and those having a negative charge are anions. An ion often has entirely different properties from the element (atom) from which it was formed.

Metal Oxidation Refining - A refining technique that removes impurities from the base metal because the impurity oxidizes more readily than the base. The metal is heated and oxygen supplied. The impurity upon oxidizing separates by gravity or volatilizes.

Metal Oxide Semiconductor Device - A metal insulator semiconductor structure in which the insulating layer is an oxide of the substrate material;l for a silicon substrate, the insulating layer is silicon dioxide ($SiO2$).

Metal Paste Production - Manufacture of metal pastes for use as pigments by mixing metal powders with mineral spirits, fatty acids and solvents. Grinding and filtration are steps in the process.

Metal Powder Production - Production of metal particles for such uses as pigments either by milling and grinding of scrap or by atomization of molten metal.

Metal Spraying - Coating metal objects by spraying molten metal upon the surface with gas pressure.

Mica - A group of aluminum silicate minerals that are characterized by their ability to split into thin, flexible flakes because of their basal cleavage.

Milligrams Per Liter (mg/l) = This is a weight per volume designation used in water and wastewater analysis.

Milling - Using a rotary tool with one or more teeth which engage the workpiece and remove material as the workpiece moves past the rotating cutter.

Mixed Media Filtration - A filter which uses two or more filter materials of differing specific gravities selected so as to produce a filter uniformly graded from coarse to fine.

Molecule - Chemical units composed of one or more atoms.

Monitoring - The measurement, sometimes continuous, of water quality.

Mount Assembly - Funnel neck ending of picture tube holding electron gun(s).

National Pollutant Discharge Elimination System (NPDES) - The Federal mechanism for regulating point source discharge by means of permits.

Navigable Waters - All navigable waters of the United States; tributaries of navigable waters of the United States; interstate waters, intrastate lakes, rivers and streams which are utilized for recreational or other purposes.

Neutralization - Chemical addition of either acid or base to a solution such as adjusting the pH to 7.

New Source - Any building, structure, facility, or installation from which there is or may be the discharge of pollutants, the construction of which is commenced after the publication of proposed regulations prescribing a standard of performance under Section 306 of the Act which will be applicable to such source if such standard is thereafter promulgated in accordance with Section 306 of the Act.

Nitrification (Biological) - The oxidation of nitrogenous matter into nitrates by bacteria.

Noble Metals - Metals below hydrogen in the electromotive force series; includes antimony, copper, rhodium, silver, gold, bismuth.

Non-Process Water - Water used for the heating and cooling of process solutions to maintain proper operating conditions or for the make-up water in cooling towers, boilers and lawn sprinkling systems. This water is not process water as it does not come in contact with raw materials or the product.

Non-Water Quality Environmental Impact - The ecological impact as a result of solid, air, or thermal pollution due to the application of various wastewater technologies to achieve the effluent guidelines limitations. Associated with the non-water quality aspect is the energy impact of wastewater treatment.

Noncontact Cooling Water - Water used for cooling which does not come into direct contact with any raw material, intermediate product, waste product, or finished product.

Nonferrous - No iron content.

Normalizing - Heat treatment of iron-base alloys above the critical temperature, followed by cooling in still air. (This is often done to refine or homogenize the grain structure of castings, forgings and wrought steel products.)

Oil-Filled Capacitor - A capacitor whose conductor and insulating elements are immersed in an insulating fluid that is usually, but not necessarily, oil.

On-Site Treatment - Treatment of effluent performed at its source, prior to discharge into a sewage system or a receiving body of water.

Organic Compound - Any substance that contains the element carbon, with the exception of carbon dioxide and various carbonates.

ORP Recorders - Oxidation-reduction potential recorders.

Outfall - The point or location where sewage or drainage discharges from a sewer, drain, or conduit.

Oxidants - Those substances which aid in the formation of oxides.

Oxidation - The conversion of chemical substances to higher oxidation states through loss of electrons. In waste treatment, oxidation usually is involved in the breakdown of many substances.

Oxide Mask - Oxidized layer of silicon wafer through which "windows" are formed which will allow for dopants to be introduced into the silicon.

Oxidizable Cyanide - Cyanide amenable to oxidation.

Oxidizing - Combining the material concerned with oxygen.

Ozonation - The process of using ozone (O3) as an oxidizing agent to oxidize and degrade chemical constituents in an effluent or to regenerate ferricyanide bleach.

Ozone - A powerful gaseous oxidizing agent (O3); it can be generated by a high voltage discharge across a stream of air or oxygen.

Paint Stripping - The term "paint stripping" shall mean the process of removing an organic coating from a workpiece or painting fixture. The removal of such coatings using processes such as caustic, acid, solvent and molten salt stripping are included.

Panel - The front, screen portion of the glass enclosure of a cathode ray tube.

Parameter - A characteristic element of constant factor.

Passivation - The changing of the chemically active surface of a metal to a much less reactive state by means of an acid dip.

Patina - A blue-green oxidation of copper.

Pearlite - A microstituent found in iron-base alloys consisting of a lamellar (Patelike) composite of ferrite and iron carbide. (This structure results from the decomposition of austenite and is very common in cast irons and annealed steels.)

Peening - Mechanical working of metal by hammer blows or shot impingement.

pH - A unit for measuring hydrogen ion concentrations. A pH of 7 indicates a "neutral water or solution. At pH lower than 7, the solution is acidic.

pH Adjustment - A means of maintaining the optimum pH through the use of chemical additives. Can be manual, automatic, or automatic with flow corrections.

pH Buffer - A substance used to stabilize the acidity or alkalinity in a solution.

Phase - One of the separate circuits or windings of a polyphase system, machine, or other apparatus.

Phase Assembly - The coil-core assembly of a single phase of a transformer.

Phenols - A group of aromatic compounds having the hydroxyl group directly attached to the benzene ring. Phenols can be a contaminant in a waste stream from a manufacturing process.

Phosphate - Salts or esters of phosphoric acid.

Phosphate Coating - Process of forming a conversion coating on iron or steel by immersing in a hot solution of manganese, iron or zinc phosphate. Often used on a metal part prior to painting or porcelainizing.

Phosphatizing - Process of forming rust-resistant coating on iron or steel by immersing in a hot solution of acid manganese, iron or zinc phosphates.

Phosphor - Crystalline inorganic compounds that produce light when excited by ultraviolet radiation.

Photolithography - The process by which a microscopic pattern is transferred from a photomask to a material layer (e.g., SiO2) in an actual circuit.

Photomask - A film or glass negative that has many high-resolution images, used in the production of semiconductor devices and integrated circuits.

Photon - A quantum of electromagnetic energy.

Photoresist - A light-sensitive coating that is applied to a substrate or board, exposed, and developed prior to chemical etching; the exposed areas serve as a mask for selective etching.

Photoresists - Thin coatings produced from organic solutions which when exposed to light of the proper wave length are chemically changed in their solubility to certain solvents (developers). This substance is placed over a surface which is to be protected during processing such as in the etching of printed circuit boards.

Photosensitive Coating - A chemical layer that is receptive to the action of radiant energy.

Pickling - The immersion of all or part of a workpiece in a corrosive media such as acid to remove scale and related surface coatings.

Picture Tube - A cathode ray tube used in television receivers to produce an image by varying the electron beam intensity as the beam scans a fluorescent screen.

Planing - Producing flat surfaces by linear reciprocal motion of the work and the table to which it is attached relative to a stationary single-point cutting tool.

Plasma Arc Machining - The term "plasma arc machining" shall mean the process of material removal or shaping of a workpiece by a high velocity jet of high temperature ionized gas.

Plate - 1. Preferably called the anode. The principal electrode to which the electron stream is attracted in an electron tube. 2. One of the conductive electrodes in a capacitor.

Plated Area - Surface upon which an adherent layer of metal is deposited.

Plating - Forming an adherent layer of metal upon an object.

Point Source - Any discernible, confined, and discrete conveyance including, but not limited to, any pipe, ditch, channel, tunnel, conduit, well, discrete fissure, container, rolling stock, concentrated animal feeding operation, or vessel or other floating craft from which pollutants are or may be discharged.

Point Source Category - See Category.

Polar Capacitor - An electrolytic capacitor having an oxide film on only one foil or electrode which forms the anode or positive terminal.

Pole Type Transformer - A transformer suitable for mounting on a pole or similar structure.

Poling - A step in the production of ceramic piezoelectric bodies which orients the oxes of the crystallites in the preferred direction.

Polishing - The process of removing stock from a workpiece by the action of loose or loosely held abrasive grains carried to the workpiece by a flexible support. Usually, the amount of stock removed in a polishing operation is only incidental to achieving a desired surface finish or appearance.

Polishing Compounds - Fluid or grease stick lubricants composed of animal tallows, fatty acids, and waxes. Selection depends on surface finish desired.

Pollutant - The term "pollutant" means dredged spoil, solid wastes, incinerator residue, sewage, garbage, sewage sludge, munitions, chemical wastes, biological materials, radioactive materials, heat, wrecked or discarded equipment, rock, sand, cellar dirt and industrial, municipal and agricultural waste discharged into water.

Pollutant Parameters - Those constituents of wastewater determined to be detrimental and, therefore, requiring control.

Pollution - The man-made or man-induced alternation of the chemical, physical, biological, and radiological integrity of water.

Pollution Load - A measure of the unit mass of a wastewater in terms of its solids or oxygen-demanding characteristics, or in terms of harm to receiving waters.

Polychlorinated Biphenyl (PCB) - A family of chlorinated biphenyls with unique thermal properties and chemical inertness which have a wide variety of uses as plasticizers, flame retardants and insulating fluids. They represent a persistent contaminant in waste streams and receiving waters.

Polyelectrolyte - A high polymer substance, either natural or synthetic, containing ionic constituents; they may be either cationic or anionic.

POTW - Publicly Owned Treatment Works (45 FR 33423).

Pouring - (Casting and Molding) Transferring molten metal from a furnace or a ladle to a mold.

Power Regulators - Transformers used to maintain constant output current for changes in temperature output load, line current, and time.

Power Transformer - Transformer used at a generating station to step up the initial voltage to high levels for transmission.

Prechlorination - 1. Chlorination of water prior to filtration. 2. Chlorination of sewage prior to treatment.

Precious Metals - Gold, silver, iridium, palladium, platinum, rhodium, ruthenium, indium, osmium, or combination thereof.

Precipitate - The discrete particles of material rejected from a liquid solution.

Precipitation - The separation of a dissolved substance from its solution by rendering it insoluble through chemical reaction.

Prehardener - A solution used to harden the emulsion in order to protect it from damage during processing.

Pressure Deformation - The process of applying force, (other than impact force), to permanently deform or shape a workpiece. Pressure deformation operations may include operations such as rolling, drawing, bending, embossing, coining, swaging, sizing, extruding, squeezing, spinning, seaming, piercing, necking, reducing, forming, crimping, coiling, twisting, winding, flaring or weaving.

Pressure Filtration - The process of solid/liquid phase separation effected by passing the more permeable liquid phase through a mesh which is impenetrable to the solid phase.

Pretreatment - Treatment of wastewaters from sources before introduction into municipal treatment works.

Primary Feeder Circuit (Substation) Transformers - These transformers (at substations) are used to reduce the voltage from the subtransmission level to the primary feeder level.

Primary Settling - The first treatment for the removal of settleable solids from wastewater which is passed through a treatment works.

Primary Treatment - The first stage in wastewater treatment in which floating or settleable solids are mechanically removed by screening and sedimentation.

Primary Winding - Winding on the supply (i.e., input) side of a transformer.

Printed Circuit Boards - A circuit in which the interconnecting wires have been replaced by conductive strips printed, etched, etc., onto an insulating board. Methods of fabrication include etched circuit, electroplating, and stamping.

Printing - A process whereby a design or pattern in ink or types of pigments are impressed onto the surface of a part.

Priority Pollutant - The 129 specific pollutants established by the EPA from the 65 pollutants and classes of pollutants as outlined in the consent decree of June 8, 1976.

Process Modification - (In-Plant Technology) Reduction of water pollution by basic changes in a manufacturing process.

Process Wastewater - Any water which, during manufacturing or processing, comes into direct contact with or results from the production or use of any raw material, intermediate product, finished product, byproduct, or waste product.

Process Water - Water prior to its direct contact use in a process or operation. (This water may be any combination of raw water, service water, or either process wastewater or treatment facility effluent to be recycled or reused).

Punching - A method of cold extruding, cold heading, hot forging or stamping in a machine whereby the mating die sections control the shape or contour of the part.

Pyrolysis - (Sludge Removal) Decomposition of materials by the application of heat in any oxygen-deficient atmosphere.

Quantity GPD - Gallons per day.

Quenching - Rapid cooling of alloys by immersion in water, oil, or gases after heating.

Raceway - A channel used to hold and protect wires, cables or busbars.

Rack Plating - Electroplating of workpieces on racks.

Racking - The placement of parts on an apparatus for the purpose of plating.

Radiography - A nondestructive method of internal examination in which metal or other objects are exposed to a beam of x-ray or gamma radiation. Differences in thickness, density or absorption, caused by internal discontinuities, are apparent in the shadow image either on a fluorescent screen or on photographic film placed behind the object.

Rapid Sandfilter - A filter for the purification of water where water which has been previously treated, usually by coagulation and sedimentation, is passed through a filtering medium consisting of a layer of sand or prepared anthracite coal or other suitable material, usually from 24 to 30 inches thick and resting on a supporting bed of gravel or a porous medium such as carborundum. The filtrate is removed by a drain system. The filter is cleaned periodically by reversing the flow of the water through the filtering medium. Sometimes supplemented by mechanical or air agitation during backwashing to remove mud and other impurities.

Raw Water - Plant intake water prior to any treatment or use.

Reaming - An operation in which a previously formed hole is sized and contoured accurately by using a rotary cutting tool (reamer) with one or more cutting elements (teeth). The principal support for the reamer during the cutting action is obtained from the workpiece. 1. Form Reaming - Reaming to a contour shape. 2. Taper Reaming - Using a special reamer for taper pins. 3. Hand Reaming - Using a long lead reamer which permits reaming by hand. 4. Pressure Coolant Reaming (or Gun Reaming) - Using a multiple-lip, end cutting tool through which coolant is forced at high pressure to flush chips ahead of the tool or back through the flutes for finishing of deep holes.

Receiving Waters - Rivers, lakes, oceans, or other water courses that receive treated or untreated wastewaters.

Recirculating Spray - A spray rinse in which the drainage is pumped up to the spray and is continually recirculated.

Reclamation - The process of deriving usable materials from waste by-products, etc., through physical or chemical treatment.

Rectifier - 1. A device for converting alternating current into direct current. 2. A nonlinear circuit component that, ideally, allows current to flow in one direction unimpeded but allows no current to flow in the other direction.

Recycle Lagoon - A pond that collects treated wastewater, most of which is recycled as process water.

Recycled Water - Process wastewater or treatment facility effluent which is recirculated to the same process.

Redox - A term used to abbreviate a reduction-oxidation reaction.

Reduction - A reaction in which there is a decrease in valence resulting from a gain in electrons.

Regeneration - The removal or reconstitution of by-products and the replacement of certain components of the processing solution being reclaimed.

Rem-Jet - A coating on the back of certain films for the reduction of light reflections during exposure. The rem-jet backing is removed during processing by washing or by washing and mechanical buffing.

Residual Chlorine - The amount of chlorine left in the treated water that is available to oxidize contaminants.

Resistor - A device designed to provide a definite amount of resistance, used in circuits to limit current flow or to provide a voltage drop.

Retention Time - The time allowed for solids to collect in a settling tank. Theoretically retention time is equal to the volume of the tank divided by the flow rate. The actual retention time is determined by the purpose of the tank. Also, the design residence time in a tank or reaction vessel which allows a chemical reaction to go to completion, such as the reduction of hexavalent chromium or the destruction of cyanide.

Reused Water - Process wastewater or treatment facility effluent which is further used in a different manufacturing process.

Reversal Process - A process which yields a direct positive image of the subject on the same material used for the original exposure.

Reverse Osmosis - The application of pressure to the surface of solution through a semipermeable membrane that is too dense to permit passage of the solute, leaving behind the dissolved solids (concentrate).

Rinse - Water for removal of dragout by dipping, spraying, fogging, etc.

Running Rinse - A rinse tank in which water continually flows in and out.

Salt - 1. The compound formed when the hydrogen of an acid is replaced by a metal or its equivalent (e.g., and NH4 radical). Example: $HCl + NaOH = NaCl + H_2O$. This is typical of the general rule that the reaction of an acid and a base yields a salt and water. Most salts ionize in water solution. 2. Common salt, sodium chloride, occurs widely in nature, both as deposits left by ancient seas and in the ocean, where its average concentration is about 3%.

Salt Bath Descaling - Removing the layer of oxides formed on some metals at elevated temperatures in a salt solution. See: Reducing, Oxidizing, Electrolytic.

Sand Blasting - The process of removing stock including surface films, from a workpiece by the use of abrasive grains pneumatically impinged against the workpiece.

Sand Filtration - A process of filtering wastewater through sand. The wastewater is trickled over the bed of sand where air and bacteria decompose the wastes. The clean water flows out through drains in the bottom of the bed. The sludge accumulating at the surface must be removed from the bed periodically.

Sanitary Sewer - A sewer that carries liquid and water wastes from residences, commercial buildings, industrial plants, and institutions together with ground, storm, and surface waters that are not admitted intentionally.

Sanitary Water - The supply of water used for sewage transport and the continuation of such effluents to disposal.

Scale - Oxide and metallic residues.

Secondary Settling Tank - A tank through which effluent from some prior treatment process flows for the purpose of removing settleable solids.

Secondary Wastewater Treatment - The treatment of wastewater by biological methods after primary treatment by sedimentation.

Sedimentation - Settling of matter suspended in water by gravity. It is usually accomplished by reducing the velocity of the liquid below the point at which it can transport the suspended material.

Semiconductor - A solid crystalline material whose electrical conductivity is intermediate between that of a metal and an insulator.

Sensitization - The process in which a substance other than the catalyst is present to facilitate the start of a catalytic reaction.

Sequestering Agent - An agent (usually a chemical compound) that "sequesters" or holds a substance in suspension.

Series Rinse - A series of tanks which can be individually heated or level controlled.

Service Water - Raw water which has been treated preparatory to its use in a process or operation; i.e., makeup water.

Settleable Solids - That matter in wastewater which will not stay in suspension during a preselected settling period, such as one hour, but either settles to the bottom or floats to the top.

Settling - The concentration of particulate matter in wastewater by allowing suspended solids to sink to the bottom.

Settling Ponds - A large shallow body of water into which industrial wastewaters are discharged. Suspended solids settle from the wastewaters due to the large retention time of water in the pond.

Sewer - A pipe or conduit, generally closed, but normally not flowing full, for carrying sewage and other waste liquids.

Shaping - Using single point tools fixed to a ram reciprocated in a linear motion past the work. 1. Form Shaping - Shaping with a tool ground to provide a specified shape. 2. Contour Shaping - Shaping of an irregular surface, usually with the aid of a tracing mechanism. 3. Internal Shaping - Shaping of internal forms such as keyways and guides.

Shaving - 1. As a finishing operation, the accurate removal of a thin layer by drawing a cutter in straight line motion across the work surfaces. 2. Trimming parts like stampings, forgings and tubes to remove uneven sheared edges or to improve accuracy.

Shearing - The process of severing or cutting of a workpiece by forcing a sharp edge or opposed sharp edges into the workpiece stressing the material to the point of sheer failure and separation.

Shipping - Transporting

Short Stop - A step in photoprocessing which follows development whereby the basic activators in the developer are neutralized to prevent further development.

Shot Peening - Dry abrasive cleaning of metal surfaces by impacting the surfaces with high velocity steel shot.

Shredding - (Cutting or Stock Removal) Material cut, torn or broken up into small parts.

SIC - Standard Industrial Classification - Defines industries in accordance with the composition and structure of the economy and covers the entire field of economic activity.

Silica - (SiO_2) Dioxide of silicon which occurs in crystalline form as quartz, crystohalite, tridymite. Used in its pure form for high-grade refractories and high temperature insulators and in impure form (i.e., sand) in silica bricks.

Silver Halide - Silver halide is an inorganic salt of silver in combination with elements from Group 7A of the Periodic Table. Silver halide salts used in photography are silver chloride, silver bromide, and silver iodide. Upon exposure to light, silver halide crystals undergo an internal change making them capable of subsequent reduction to metallic silver by appropriate developing agents.

Silvering - The deposition of thin films of silver on glass, etc. carried by one of several possible processes.

Silver Recovery - Removal of silver from used photographic processing solutions and materials so it can be made available for reuse.

Sintering - The process of forming a mechanical part from a powdered metal by bonding under pressure and heat but below the melting point of the basis metal.

Skimming - The process of removing floating solid or liquid wastes from a wastewater stream by means of a special tank and skimming mechanism prior to treatment of the water.

Slaking - The process of reacting lime with water to yield a hydrated product.

Sludge - Residue produced in a waste treatment process.

Sludge Cake - The material resulting from air drying or dewatering sludge (usually forkable or spadable).

Sludge Dewatering - The removal of water from sludge by introducing the water sludge slurry into a centrifuge. The sludge is driven outward with the water remaining near the center. The water is withdrawn and the dewatered sludge is usually landfilled.

Sludge Disposal - The final disposal of solid wastes.

Sludge Thickening - The increase in solids concentration of sludge in sedimentation or digestion tank.

Slurry - A watery suspension of solid materials.

Soldering - The process of joining metals by flowing a thin (capillary thickness) layer of nonferrous filler metal into the space between them. Bonding results from the intimate contact produced by the dissolution of a small amount of base metal in the molten filler metal, without fusion of the base metal.

Solids - (Plant Waste) Residue material that has been completely dewatered.

Solute - A dissolved substance.

Solution - Homogeneous mixture of two or more components such as a liquid or a solid in a liquid.

Solvent - A liquid capable of dissolving or dispersing one or more other substances.

Solvent Cleaning - Removal of oxides, soils, oils, fats, waxes, greases, etc. by solvents.

Solvent Degreasing - The removal of oils and grease from a workpiece using organic solvents or solvent vapors.

Specific Conductance - The property of a solution which allows an electric current to flow when a potential difference is applied.

Spectrophotometry - A method of analyzing a wastewater sample by means of the spectra emitted by its constituents under exposure to light.

Spray Rinse - A process which utilizes the expulsion of water through a nozzle as a means of rinsing.

Spray Washing - A method of washing film or paper using a spray rather than an immersion tank as a means of conserving water.

Sputtering - A process to deposit a thin layer of metal on a solid surface in a vacuum. Ions bombard a cathode which emits the metal atoms.

Squeegee - A piece of flexible material or a thin stream of air set to impinge on one or both sides of photographic film or paper as it comes out of a tank of processing solution. This reduces the amount of solution carried over.

Stabilizer - A chemical bath, usually the last in a processing cycle, that imparts greater life to a processed photographic film or paper through one of several preserving steps.

Stacked Capacitor - Device containing multiple layers of dielectric and conducting materials and designed to store electrical charge.

Stainless Steels - Steels which have good or excellent corrosion resistance. (One of the common grades contains 18% chromium and 8% nickel.) There are three broad classes of stainless steels - ferritic, austenitic, and martensitic. These various classes are produced through the use of various alloying elements in differing quantities.

Stamping - Almost any press operations including blanking, shearing, hot and cold forming, drawing, blending, or coining.

Standard of Performance - Any restrictions established by the Administrator pursuant to Section 306 of the Act on quantities, rates and concentrations of chemical, physical, biological, and other constituents which are or may be discharged from new sources into navigable waters, the waters of the contiguous zone or the ocean.

Stannous Salt - Tin based compound used in the acceleration process. Usually stannous chloride.

Steel - An iron-based alloy, malleable under proper conditions, containing up to about 2% carbon.

Step-Down Transformers - (Substation) - A transformer in which the AC voltages of the secondary windings are lower than those applied to the primary windings.

Step-Up Transformers - Transformer in which the energy transfer is from a low-voltage primary (input) winding to a high-voltage secondary (output) winding or windings.

Storm Water Lake - Reservoir for storage of storm water runoff collected from plant site; also, auxiliary source of process water.

Stress Relieved - The heat treatment used to relieve the internal stresses induced by forming or heat treating operations. (It consists of heating a part uniformly, followed by cooling slow enough so as not to reintroduce stresses. To obtain low stress levels in steels and cast irons, temperatures as high as 1250 degrees F may be required.)

Strike - A thin coating of metal (usually less than 0.0001 inch in thickness) to be followed by other coatings.

Stripping - The removal of coatings from metal.

Studs - Metal pins in the glass of picture tubes onto which shadow mask is hung.

Subcategory or Subpart - A segment of a point source for which specific effluent limitations have been established.

Substation - Complete assemblage of plant, equipment, and the necessary buildings at a place where electrical energy is received (from one or more power stations) for conversion (e.g., from AC to DC by means of rectifiers or rotary converters), for stepping-up or down by means of transformers, or for control (e.g., by means of switch-gear, etc.).

Substrates - Thin coatings (as of hardened gelatin) which act as a support to facilitate the adhesion of a sensitive emulsion.

Subtractive Circuitry - Circuitry produced by the selective etching of a previously deposited copper layer.

Surface Tension - A measure of the force opposing the spread of a thin film of liquid.

Surface Waters - Any visible stream or body of water.

Surfactants - Surface active chemicals which tend to lower the surface tension between liquids, such as between acid and water.

Surge - A sudden rise to an excessive value, such as flow, pressure, temperature.

Suspended Solids - Undissolved matter carried in effluent that may settle out in a clarifier.

Swaging - Forming a taper or a reduction on metal products such as rod and tubing by forging, squeezing or hammering.

Tank - A receptacle for holding transporting or storing liquids.

Tantalum - A lustrous, platinum-gray ductile metal used in making dental and surgical tools, penpoints, and electronic equipment.

Tantalum Foil - A thin sheet of tantalum, usually less than 0.006 inch thick.

Tapping - Producing internal threads with a cylindrical cutting tool having two or more peripheral cutting elements shaped to cut threads of the desired size and form. By a combination of rotary and axial motion, the leading end of the tap cuts the thread while the tap is supported mainly by the thread it produces.

Tempering - Reheating a quench-hardened or normalized ferrous alloy to a temperature below the transformation range then cooling at any rate desired.

Terminal - A screw, soldering lug, or other point to which electric connections can be made.

Testing - A procedure in which the performance of a product is measured under various conditions.

Thermoplastic Resin - A plastic that solidifies when first heated under pressure, and which cannot be remelted or remolded without destroying its original characteristics; examples are epoxides, melamines, phenolics and ureas.

Thickener - A device or system wherein the solid contents of slurries or suspensions are increased by gravity settling and mechanical separation of the phases, or by flotation and mechanical separation of the phases.

Thickening - (Sludge Dewatering) Thickening or concentration is the process of removing water from sludge after the initial separation of the sludge from wastewater. The basic objective of thickening is to reduce the volume of liquid sludge to be handled in subsequent sludge disposal processes.

Threading - Producing external threads on a cylindrical surface. 1. Die Threading - A process for cutting external threads on cylindrical or tapered surfaces by the use of solid or self-opening dies. 2. Single-Point Threading - Turning threads on a lathe. 3. Thread Grinding - See definition under grinding. 4. Thread Milling - A method of cutting screw threads with a milling cutter.

Threshold Toxicity - Limit upon which a substance becomes toxic or poisonous to a particular organism.

Through Hole Plating - The plating of the inner surfaces of holes in a printed circuit board.

Titration - 1. A method of measuring acidity or alkalinity. 2. The determination of a constituent in a known volume of solution by the measured addition of a solution of known strength for completion of the reaction as signaled by observation of an end point.

Total Chromium - The sum of chromium in all valences.

Total Cyanide - The total content of cyanide expressed as the radical CN- or alkali cyanide whether present as simple or complex ions. The sum of both the combined and free cyanide content of a plating solution. In analytical terminology, total cyanide is the sum of cyanide amenable to oxidation by chlorine and that which is not according to standard analytical methods.

Total Dissolved Solids (TDS) - The total amount of dissolved solid materials present in an aqueous solution.

Total Metal - Sum of the metal content in both soluble and insoluble form.

Total Organic Carbon (TOC) - TOC is a measure of the amount of carbon in a sample originating from organic matter only. The test is run by burning the sample and measuring the CO_2 produced.

Total Solids - The sum of dissolved and undissolved constituents in water or wastewater, usually stated in milligrams per liter.

Total Suspended Solids (TSS) - Solids found in wastewater or in the stream, which in most cases can be removed by filtration. The origin of suspended matter may be man-made or of natural sources, such as silt from erosion.

Total Volatile Solids - Volatile residue present in wastewater.

Toxic Pollutants - A pollutant or combination of pollutants including disease causing agents, which after discharge and upon exposure, ingestion, inhalation or assimilation into any organism either directly or indirectly cause death, disease, cancer, genetic mutations, physiological malfunctions (including malfunctions in such organisms and their offspring).

Transformer - A device used to transfer electric energy, usually that of an alternating current, from one circuit to another; especially, a pair of multiply-wound, inductively coupled wire coils that effect such a transfer with a change in voltage, current, phases, or other electric characteristics.

Transistor - An active component of an electronic circuit consisting of a small block of semiconducting material to which, at least three electrical contacts are made; used as an amplifier, detector, or switch.

Treatment Facility Effluent - Treated process wastewater.

Trepanning - Cutting with a boring tool so designed as to leave an unmachined core when the operation is completed.

Trickling Filter - A filter consisting of an artificial bed of course material, such as broken stone, clinkers, slats, or brush over which sewage is distributed and applied in drops, films, or spray, from troughs, drippers, moving distributors or fixed nozzles and through which it trickles to the underdrain giving opportunity for the formation of zoogleal slimes which clarify the oxidized sewage.

Turbidimeter - An instrument for measurement of turbidity in which a standard suspension is usually used for reference.

Turbidity - 1. A condition in water or wastewater caused by the presence of suspended matter resulting in the scattering and absorption of light rays. 2. A measure of fine suspended matter in liquids. 3. An analytical quantity usually reported in arbitrary turbidity units determined by measurements of light diffraction.

Turning - Generating cylindrical forms by removing metal with a single-point cutting tool moving parallel to the axis of rotation of the work. 1. Single-Point Turning - Using a tool with one cutting edge. 2. Face Turning - Turning a surface perpendicular to the axis of the workpiece. 3. Form Turning - Using a tool with a special shape. 4. Turning Cutoff - Severing the workpiece with a special lathe tool. 5. Box Tool Turning - Turning the end of the workpiece with one or more cutters mounted in a boxlike frame, primarily for finish cuts.

Ultrafiltration - A process using semipermeable polymeric membranes to separate molecular or colloidal materials dissolved or suspended in a liquid phase when the liquid is under pressure.

Ultrasonic Cleaning - Immersion cleaning aided by ultrasonic waves which cause microagitation.

Ultrasonic Machining - Material removal by means of an ultrasonic-vibrating tool usually working in an abrasive slurry in close contact with a workpiece or having diamond or carbide cutting particles on its end.

Unit Operation - A single, discrete process as part of an overall sequence, e.g., precipitation, settling and filtration.

Vacuum Deposition - Condensation of thin metal coatings on the cool surface of work in a vacuum.

Vacuum Evaporization - A method of coating articles by melting and vaporizing the coating material on an electrically heated conductor in a chamber from which air has been exhausted. The process is only used to produce a decorative effect. Gold, silver, copper and aluminum have been used.

Vacuum Filter - A filter consisting of a cylindrical drum mounted on horizontal axis, covered with a filter cloth revolving with a partial submergence in liquid. A vacuum is maintained under the cloth for the larger part of a revolution to extract moisture and the cake is scraped off continuously.

Vacuum Filtration - A sludge dewatering process in which sludge passes over a drum with a filter medium, and a vacuum is applied to the inside of the drum compartments. As the drum rotates, sludge accumulates on the filter surface, and the vacuum removes water.

Vacuum Metalizing - The process of coating a workpiece with metal by flash heating metal vapor in a high-vacuum chamber containing the workpiece. The vapor condenses on all exposed surfaces.

Vacuum Tube - An electron tube vacuated to such a degree that its electrical characteristics are essentially unaffected by the presence of residual gas or vapor.

Vapor Blasting - A method of roughing plastic surfaces in preparation for plating.

Vapor Degreasing - Removal of soil and grease by a boiling liquid solvent, the vapor being considerably heavier than air. At least one constituent of the soil must be soluble in the solvent.

Vapor Plating - Deposition of a metal or compound upon a heated surface by reduction or decomposition of a volatile compound at a temperature below the melting points of either the deposit or the basis material.

Variable Capacitor - A device whose capacitance can be varied continuously by moving one set of metal plates with respect to another.

Volatile Substances - Material that is readily vaporizable at a relatively low temperature.

Voltage Breakdown - The voltage necessary to cause insulation failure.

Voltage Regulator - Like a transformer, it corrects changes in current to provide continuous, constant current flow.

Volumetric Method - A standard method of measuring settleable solids in an aqueous solution.

Wash - A water wash is a step in photoprocessing removing residual processing chemicals absorbed in the emulsion or substrate.

Waste Discharged - The amount (usually expressed as weight) of some residual substance which is suspended or dissolved in the plant effluent.

Wastewater Constituents - Those materials which are carried by or dissolved in a water stream for disposal.

Welding - The process of joining two or more pieces of material by applying heat, pressure of both, with or without filler material, to produce a localized union through fusion or recrystallization across the interface.

Wet Air Scrubber - Air pollution control device which uses a liquid or vapor to absorb contaminants and which produces a wastewater stream.

Wet Slug Capacitor - Refers to a sintered tantalum capacitor where the anode is placed in a metal can, filled with an electrolyte and then sealed.

Wet Tantalum Capacitor - A polar capacitor the cathode of which is a liquid electrolyte (a highly ionized acid or salt solution).

Dictionary of Acronyms

API - American Petroleum Institute
BLIX - Bleach-Fix (Solution)
BOD - Biochemical Oxygen Demand
BPT - Best Practical Technology
CDTA - A Chelating Agent in the Amino Carboxylic Acid Group
CFR - Code of Federal Regulations
COD - Chemical Oxygen Demand
CRC - Chemical Recovery Cartridge
CRT - Cathode Ray Tubes
CWA - Clean Water Act of 1972
CWF - Combined Wastestream Formula
DAF - Dissolved Air Flotation
DC - Color in Developer
DHEG - A Chelating Agent in the Amino Carboxylic Acid Group
DI - De-ionized
DOHS - Department Of Health Services
DPDEED - A Chelating Agent in the Amino Carboxylic Acid Group
DTA - A Chelating Agent in the Amino Carboxylic Acid Group
DTPA - A Chelating Agent in the Amino Carboxylic Acid Group
EDTA - Ethylenediamine Tetraacetic Acid
EGTA - A Chelating Agent in the Amino Carboxylic Acid Group
EMR - Electrolytic Metal Recovery
EPA - Environmental Protection Agency
FOG - Fats, Oils and Grease
GCMS - Gas Chromatograph Mass Spectrophotometry
GPD - Gallons Per Day
GPM - Gallons Per Minute
HEDTA - A Chelating Agent in the Amino Carboxylic Acid Group
HEIDA - A Chelating Agent in the Amino Carboxylic Acid Group
IAP - Induced Air Flotation
IC - Color Incorporated
ISD - Interim Status Document
IW - Industrial Waste
LCD - Liquid Crystal Display
LED - Light Emitting Diodes
MOS - Metal Oxide Silicon

MSDS - Material Safety Data Sheet
N/A - Not Applicable
NPDES - National Pollutant Discharge Elimination System
NRDC - Natural Resources Defense Council
PCB - Printed Circuit Board (Also known as PC Board)
PL - Public Law
POTW - Publicly Owned Treatment Works
PPDT - A Chelating Agent in the Amine Group
PPI - Parallel Plate Interceptors
PSES - Pretreatment Standards for Existing Sources
PSI - Pounds per Square Inch
PSNS - Pretreatment Standards for New Sources
RO - Reverse Osmosis
SIC - Standard Industrial Classification
TBED - A Chelating Agent in the Amine Group
TDS - Total Dissolved Solids
TEA - Triethanolamine
TPA - A Chelating Agent in the Amine Group
TPED - A Chelating Agent in the Amine Group
TSDF - Transfer, Storage and Disposal Facility
TSS - Total Suspended Solids
TTO - Total Toxic Organics

Index

Abrasive jet machining, L-66
Acetone, labs, D-3
Acids, L-24, 56
Acid cleaning, L-15, 51
Activated carbon adsorption, M-10
Additive production method, L-48
Adsorption, G-4
Agricultural-landscaping, horticulture, K-2
Alkaline cleaning, L-15, 51
Alkaline hydrolysis, M-4
Alkaline precipitation, M-7
Alpha radiation, C-7
Alternate mass limit formula, L-79
Aluminum, L-19, 33
Ammonium persulfate, L-40
Anodizing, L-28
Assembly, L-70
Auto
   new car dealerships, A-2
   mechanics, K-1
   rental agencies, A-3
   repair services, A-1, 6
   repair shops, A-6, 8

Barrel finishing, L-63
Batch discharge, L-73
Benzene, D-3, G-4, 7, 11
Beta radiation, C-7
Biological and natural extraction products, G-2, 3, 6, 11
Black and white processing, H-1, 3
Blowdown, J-1, 6
BOD, B-1, G-4, 7, 8, 10
Body shops, A-27
Boilers
   fire tube, J-3
   hospital, C-11
   industrial, J-1
   laundries, E-6
   plants, K-1
   water tube, J-3
Brass, L-19, 33
Brazing, L-64
Briny water, J-9
Bronze, L-19, 33
Burnishing, L-63

Cadmium, L-5, 6a, 6b, 19, 33, O-2a, 2b
Calibration, L-71
Captive printing operations, I-6
Car washes, A-26, K-1
Carpentry shops, K-2
Catalyst application, L-52

Categorical, L-1
Categorical pretreatment standards
   electroplaters, L-4
   metal finishers, L-6
Cathode ray tubes (CRT) limitations, O-1
   PSES, O-2a
   PSNS, O-2b
Caustic, A-17, 18
Central service/supply, hospital, C-9
Chelating agents, L-24, 57
Chemical chromium reduction, M-6
Chemical milling and etching, L-40
Chemical synthesis, G-11
Chemical synthesis products, G-2, 3
Chlorination, M-3
Chloroform, G-4, 11
Chromic acid, L-40
Chromium, F-3, H-8, L-5, 6a, 18, 19, 51, O-1, 2a, 2b
Chromium total, L-6b
Clarification, M-11
Clarifiers laundries, E-8
Cleaning and surface preparations, L-51
Cleaning, L-62
Coatings, L-33
COD, G-4, 7, 8, 10
Color processing, H-1, 4, 6
Combined waste steam formula, L-76–78
Combustibles, labs, D-3
Compliance dates, L-80
Condensate, J-1, 8
Control technology laundries, E-3
Cooling, water system/chillers, C-11, E-6
Cooling towers, H-10
Cooling systems, Q-1
   open recirculating, Q-2
   closed recirculating, Q-3
   one pass, Q-4
   cooling towers, Q-5
Copper, A-17, 18, F-3, G-5, L-5, 6a, 6b, 19, 33, 51, 53, 56, O-1
Crystallization, M-17
Cupric chloride, L-40
Cyanide total, L-6a, 6b
Cyanide, G-12, H-7, I-5, 9, L-5, 6c, 18, 54, 57
Cyanide destruction, M-3
Cyclohexane, G-7

Decreasing, L-15
Demineralization, J-9

Descaling, L-16
Detail shops, A-27
Dicloroethylene, G-4
Dining/and eating facilities, K-2
Direct precipitation, G-4
Dissolved air flotation, E-10
Dopants, P-3
Drag out reduction, M-26

Electrical discharge machining, L-66
Electrochemical machining, L-66
Electrodialysis, M-18
Electroless plating, L-23, 56
Electroless copper plate, L-53
Electrolytic metal recovery, M-14
Electron tubes, O-2
Electron beam machining, L-66
Electroplating overview, L-8, 9, 12
   anodizing, L-12
   chemical etching, L-12
   coating, L-12
   electroless plating, L-12
   printed circuit board manufacturing, L-12
Electroplating, L-19, 54, P-4
Electroplating limitations, L-5
Emulsion cleaning, L-15
Engine repair shops, A-11
EPA, E-7
EPA discharge limits, H-2, 13
Equalization, E-9, M-22
Equalization pits laundries, E-8
Equalization tanks food processing, B-1
Etching/surface preparation, L-51, 56
Ether, D-3
Ethylene oxide, C-9
Ethylene-glycol, A-17
Evaporation, M-11

Fermentation, G-11
Fermentation products, G-2, 3
Ferric chloride, L-40
Ferrous sulfate, M-6
Filter settling, A-19
Filtration, M-11
Fire fighting systems, E-6
Flame spraying, L-65
Flammables, D-3
Flotation, M-11
Fluoborate, L-54
Fluoride, L-57, O-2a, b
Food processing, B-1
Formaldehyde, L-53
Formulation products, G-2, 3

Gallium, C-7
Gold, L-19, 57, O-1
Grease and oil, C-9, E-2
Grease and oil control systems, C-10, M-23
Grinding, L-62

Hazardous waste disposal/storage, C-12
Hazardous waste storage/treatment facilities, K-2
Heat treating, L-64
Heavy metals, E-2
High pressure temperature technique, M-4
High pressure high temperature washing, A-24
Hospitals, C-1–C-15
Hot dip coating, L-68
Hydraulic fluids, A-6

Immersion plating, L-34, 56
Impact deformation, L-63
Indium, L-19
Institutions, K-1
Iodine (123), C-7
Ion exchange, G-4, 5, M-16
Iridium, L-19
Iron, L-19

Kitchen/cafeteria, hospital, C-10

Laboratories, K-2
Laboratories, medical/clinical, D-1
  bacteriology/microbiology, C-4
  chemistry, D-1
  clinical, pathology, bacteriology, C-3
  commercial, D-1
  cytology, C-4
  histology, C-4
  pathology/histology, D-1
  research, D-1
  toxicology, C-4, D-1
  X-ray, D-1
Lamination, L-68
Laser beam machining, L-67
Laun water use laundries, E-5
Laundries
  central, services hospital, C-9
  commercial and coin-operated, E-1, K-2
Lead, A-17, 18, E-8, F-3, L-5, 6a, 6b, 19, 57, N-7, O-2a, 2b
Local limits, L-7

Machine shops, A-6
Machining, L-62
Magnesium, L-33
Marble chip neutralization, D-3
Mechanical plating, L-71
Mercury, F-3
Metal coloring, L-34

Metal finishing, L-8, 10, 13, K-1
Metal finishing/electroplating, L-1
Metal finishing limitations
  PSES, L-6a
  PSNS, L-6b
Metal finishing process operations, additional 40, L-62
Metal working, K-1
Metal cleaning, A-7
Metals removal, M-6
Military installations, K-1
Mini labs, H-2, 12
Morgue, hospital, C-6

Negative developing, I-4
Nickel, L-5, 6a, 6b, 18, 19, 56, O-1
Nickel tab plating, L-47
Nitric acid, L-40
Nuclear medicine, C-7

Off-set printing, I-4
Oil and grease, E-9, N-7
Oil recovery, N-1, 3
  reclamation, N-3
  recycling, N-3
  reprocessing, N-3
Oils, lubricating, quenching, cutting, A-6
Oily waste pretreatment, N-1, 5
  air flotation, N-5
  emulsion breaking, N-5
  gravity separation, N-5
  ultrafiltration, N-6
Organic solvents laundries, E-2
Osmium, L-19
Oxidation/ozonation, M-21
Ozonation, M-4

Paint
  electro, L-69
  electrostatic, L-69
  ink formulation, F-1
  painting, L-69
  shops, A-27, K-2
  stripping, L-69
Palladium, L-19
Panel or pattern plating, L-46, 56
Paper, I-1
Passivating, L-33
PCBs, L-55
Petroleum distillates, I-7
pH, D-3, G-7, 8, 10, L-18, N-7
  food processing, B-1
  laundries, E-4, 8
  neutralization, M-19
Pharmaceutical, G-1
Phenol, G-5
Phosphating, L-33
Phosphorescent coatings, O-1
Phosphorus, L-57
Photo finishing, H-1

Photo printing graphic arts, C-11, K-2
Physical plants, K-1
Physical therapy, hospital, C-8
Plasma arc machining, L-67
Plate exposure/processing, I-4
Plating application, L-17
Platinum, L-19
Polishing, L-63
Post treatment, L-18
Pottery–ceramics, jewelry making, K-2
Pressure deformation, L-63
Pretreatment
  laundries, E-9
  regulations, L-3
  silver recovery, H-2, 14–16, I-10
Printed circuit boards, L-45
Printing/publishing, I-1
Process solutions, L-72
Processing equipment, H-2, 11
Pyridine, G-7
Pyrophosphate, L-54

Radiator shops, A-15–23
  flushing, A-15, 16, 17
Raw materials and products, G-11
Regeneration, J-6
Reverse osmosis, C-11, M-15
Rhodium, L-19, O-1
Rinse tanks, L-74
Rinsing techniques, M-24
  closed loop, M-24
  combination, M-27
  dead, still, M-25
  series, M-24
  single running, M-25
  spray, M-25
Ruthenium, L-19

Salt bath descalling, L-69
Sand blasting, L-65
Scrubber water, J-9
Scrubbers, G-10, L-51
Sealing operations, L-35
Semi-additive production method, L-49
Semiconductor limitations
  PSES, P-5b
Shearing, L-63
Silicon-based integrated circuits, P-5
Silkscreen printing, I-3
Silkscreening, L-56
Silver, H-5, 6, 14, 16, I-9, L-5, 6a, 6b, 19, 33, O-1
Single pass cooling, D-2
Sintering, L-67
Skimming, laundries, E-9
Sludge, A-18, E-10, L-73
Sodium persulfate, L-40
Sodium hydroxide, L-40, 53
Sodium metabisulfite, M-6
Sodium bisulfite, M-6

Soft water production, hospital, C-11
Softening, J-6
Soldering, L-65
Solids removal, M-11
Solvent extraction, G-4
Solvent recovery and removal, M-9
Solvent decreasing, L-69
Solvents, G-6, I-1, 6, 7, L-15
Sources of water pollution, L-72
Spent solvents, F-1
Sputtering, L-68, P-4
Steam cleaning, A-24, K-1
Stem plants, J-1, 2
Substrative production, L-46
Sulfate, L-54
Sulfide precipitation, M-8
Sulfur dioxide, M-6
Surface cleaning, L-56
Surface preparation, L-15

Technesium (99), C-7
Testing, L-71
Thermal infusion, L-68
Thermal cutting, L-64
Tin, L-19
Titanium, F-3
Toluene, G-7
Total metals, L-5
Total suspended solids, G-4, 7, 8, 10, N-7
Toxic organics, L-5, 6a, 6b, 57, O-2a, 2b
Toxic solvents, D-3
Transfer procedure/processes, I-5
Transmission repair shops, A-11
Treatment technologies, M-1

Ultrasonic machining, L-67
Ultrasonic cleaning, L-15
University, K-1

Vacuum deposition, P-4
Vacuum metalizing, L-70
Vapor plating, L-68
Vehicle washing/transportation, hospital, C-11

Waste minimization, L-74
Water use, H-2, 9
Water treatment, J-1, 5
Water conservation, L-74
Watery paint wastes, F-1
Welding, L-64

X-ray department and nuclear medicine, hospital, C-7
Xylene, C-4, D-3, G-7

Zinc, A-17, 18, E-8, G-5, N-7, L-5, 6a, 6b, 18, 19, 33, O-2a, 2b